Cancer Sensitizing Agents for Chemotherapy

EPIGENETIC REGULATION IN OVERCOMING CHEMORESISTANCE

VOLUME 16

Cancer Sensitizing Agents for Chemotherapy Series

Series Editor: Benjamin Bonavida, PhD

Volume 1: Nitric Oxide (Donor/Induced) in Chemosensitization
Edited by Benjamin Bonavida, PhD

Volume 2: Role of Nutraceuticals in Cancer Chemosensitization
Edited by Alok Chandra Bharti and Bharat Bhushan Aggarwal

Volume 3: Targeting Cell Survival Pathways to Enhance Response to Chemotherapy
Edited by Daniel E. Johnson, PhD

Volume 4: Protein Kinase Inhibitors as Sensitizing Agents for Chemotherapy
Edited by Zhe-Sheng Chen and Dong-Hua Yang

Volume 5: Breaking Tolerance to Pancreatic Cancer Unresponsiveness to Chemotherapy
Edited by Ganji Purnachandra Nagaraju, PhD, DSc

Volume 6: Improving the Therapeutic Ratio in Head and Neck Cancer
Edited by Randall J. Kimple, MD, PhD

Volume 7: Drug Efflux Pumps in Cancer Resistance Pathways: From Molecular Recognition and Characterization to Possible Inhibition Strategies in Chemotherapy
Edited by Alejandro Sosnik, PhD and Reina Bendayan, PharmD

Volume 8: Drug Resistance in Colorectal Cancer: Molecular Mechanisms and Therapeutic Strategies
Edited by Chi Hin Cho and Tao Hu, PhD

Volume 9: Novel Therapies in Head and Neck Cancer: Beyond the Horizon
Edited by Maie A. St. John, MD, PhD, FACS and No-Hee Park, DMD, PhD

Volume 10: pH-Interfering Agents as Chemosensitizers in Cancer Therapy
Edited by Claudiu T. Supuran, PhD and Simone Carradori, PhD

Volume 11: Overcoming Ovarian Cancer Chemoresistance
Edited by Goli Samimi, PhD, MPH and Christina Annunziata, MD, PhD

Volume 12: Biological Mechanisms and the Advancing Approaches to Overcoming Cancer Drug Resistance
Edited by Andrew Freywald, PhD and Franco J. Vizeacoumar, PhD

Volume 13: Therapeutic Strategies to Overcome ALK Resistance in Cancer
Edited by Luc Friboulet, PhD

Volume 14: New Targeting in the Reversal of Resistant Glioblastomas
Edited by Ali Syed Arbab, MD, PhD

Volume 15: Glioblastoma Resistance to Chemotherapy: Molecular Mechanisms and Innovative Reversal Strategies
Edited by Ramasamy Paulmurugan, PhD and Tarik F. Massoud, MD, PhD

Volume 16: Epigenetic Regulation in Overcoming Chemoresistance
Edited by Chunfu Wu, PhD and Lihui Wang, PhD

Volume 17: Overcoming Drug Resistance in Gynecologic Cancers
Edited by Riyaz Basha, PhD, FAACC, FABAP and Sarfraz Ahmad, PhD, FAACC, FABAP

Upcoming Volumes:

Terpenoids as Chemosensitizers Against Drug-resistant Tumors
Edited by Thomas Efferth

Chemotherapy Sensitizing Agents for Prostate Cancer
Edited by K. Ketola

Chemosensitization by TRK Inhibitors
Edited by Dirk Hempel

Chemosensitization by BTK Inhibitors
Edited by Varinder Kaur

Chemosensitization by Anti-angiogenic Agents
Edited by Lucia Morbidelli

Chemosensitization by IDH1 Inhibitors
Edited by Robert Musiol

Chemosensitization by EGFR Inhibitors
Edited by M. Pancione

Chemosensitization by IDH2 Inhibitors
Edited by Sanjee Singh

Chemosensitization by Bcl2 Inhibitors
Edited by Yu Yu

Chemosensitization by BTK Inhibitors (II)
Edited by Jean Koff

Overcoming Resistance to EGFR Inhibitors in EGFR Mutant NSCLC
Edited by Anthony Faber

Cancer Sensitizing Agents for Chemotherapy

EPIGENETIC REGULATION IN OVERCOMING CHEMORESISTANCE

VOLUME 16

Edited by

CHUNFU WU, PHD
Department of Pharmacology, Shenyang Pharmaceutical University, Shenyang, People's Republic of China

LIHUI WANG, PHD
Department of Pharmacology, Shenyang Pharmaceutical University, Shenyang, People's Republic of China

ELSEVIER

ACADEMIC PRESS
An imprint of Elsevier

Academic Press is an imprint of Elsevier
125 London Wall, London EC2Y 5AS, United Kingdom
525 B Street, Suite 1650, San Diego, CA 92101, United States
50 Hampshire Street, 5th Floor, Cambridge, MA 02139, United States
The Boulevard, Langford Lane, Kidlington, Oxford OX5 1GB, United Kingdom

Notices
Knowledge and best practice in this field are constantly changing. As new research and experience broaden our understanding, changes in research methods, professional practices, or medical treatment may become necessary.

Practitioners and researchers must always rely on their own experience and knowledge in evaluating and using any information, methods, compounds, or experiments described herein. In using such information or methods they should be mindful of their own safety and the safety of others, including parties for whom they have a professional responsibility.

To the fullest extent of the law, neither the Publisher nor the authors, contributors, or editors, assume any liability for any injury and/or damage to persons or property as a matter of products liability, negligence or otherwise, or from any use or operation of any methods, products, instructions, or ideas contained in the material herein.

Library of Congress Cataloging-in-Publication Data
A catalog record for this book is available from the Library of Congress

British Library Cataloguing-in-Publication Data
A catalogue record for this book is available from the British Library

ISSN: 2468-3183

ISBN: 978-0-12-823684-0

For information on all Academic Press publications
visit our website at https://www.elsevier.com/books-and-journals

Publisher: Stacy Masucci
Senior Acquisitions Editor: Rafael Teixeira
Editorial Project Manager: Samantha Allard
Production Project Manager: Punithavathy Govindaradjane
Cover Designer: Miles Hitchen

Typeset by SPi Global, India

Cover Image Insert

The dysregulation of epigenetic regulation including DNA methylation, histone modification, microRNA, and lncRNA is a machinery that leads to chemoresistance. Epigenetic agents targeting epigenetic modulators are potential chemotherapy sensitizing agents that have shown property of reversing drug resistance in both basic and clinical trials.

Aims and Scope for Series "Cancer Sensitizing Agents for Chemotherapy"

Current cancer management strategies fail to adequately treat malignancies with chemotherapy, with multivariable dose-restrictive factors such as systemic toxicity and multidrug resistance; hence, limiting therapeutic benefits, quality of life, and complete long-term remission rates. The resistance of cancer cells to anticancer drugs is one of the major reasons for the failure of traditional cancer treatments. Cellular components and dysregulation of signaling pathways contribute to drug resistance. If modulated, such perturbations may restore the drug response and its efficacy. The recent understanding of the molecular mechanisms and targets that are implicated for cancer chemoresistance has paved the way to develop a large battery of small molecules (sensitizing agents) that can target resistant factors and reduce the threshold of resistance, thus allowing their combination with chemotherapeutic drugs to be effective and to reverse the chemoresistance. A large variety of chemotherapy-sensitizing agents has been developed and several have been shown to be effective in experimental models and in cancer patients.

The main objective of the proposed series "Cancer Sensitizing Agents for Chemotherapy" is to publish individualized and focused volumes whereby each volume is edited by an invited expert Editor(s). Each volume will dwell on specific chemosensitizing agents with similar targeting activities. The combination treatment of the sensitizing agent with chemotherapy may result in a synergistic/additive activity and the reversal of tumor cells resistance to drugs.

The Editor(s) will compile nonoverlapping review chapters on reported findings in various cancers, both experimentally and clinically, with particular emphases on underlying biochemical, genetic, and molecular mechanisms of the sensitizing agent and the combination treatment.

The scope of the series is to provide scientists and clinicians with updated basic and clinical information that will be valuable in their quest to investigate, develop, and apply novel combination therapies to reverse drug resistance, thereby prolonging survival and even resulting in a cure in cancer patients.

Dr. Benjamin Bonavida, PhD (Series Editor)

About the Series Editor

Dr. Benjamin Bonavida, PhD, (Series Editor), is currently Distinguished Research Professor at the University of California, Los Angeles (UCLA). He is affiliated with the Department of Microbiology, Immunology and Molecular Genetics, UCLA David Geffen School of Medicine. His research career, thus far, has focused on investigations in the fields of basic immunochemistry and cancer immunobiology. His research investigations have ranged from the biochemical, molecular, and genetic mechanisms of cell-mediated killing and tumor cell resistance to chemo-immuno cytotoxic drugs. The reversal of tumor cell resistance was investigated by the use of various selected sensitizing agents based on molecular mechanisms of resistance. In these investigations, there was the newly characterized dysregulated NF-κB/Snail/YY1/RKIP/PTEN loop in many cancers that was reported to regulate cell survival, proliferation, invasion, metastasis, and resistance. Emphasis was focused on the roles of the tumor suppressor Raf Kinase Inhibitor Protein (RKIP), the tumor promoter Yin Yang 1 (YY1), and the role of nitric oxide as a chemo-immuno-sensitizing factor. Many of the earlier mentioned studies are centered on the clinical challenging features of cancer patients' failure to respond to both conventional and targeted therapies.

The Editor has been active in the organization of regular sequential international miniconferences that are highly focused on the roles of YY1, RKIP, and nitric oxide in cancer and their potential therapeutic applications. Several books edited or coedited by the Editor have been published. In addition, the Editor has been the Series Editor of books (over 23) published by Springer on *Resistance to Anti-Cancer Targeted Therapeutics*. In addition, the Editor is presently the Series Editor of three series published by Elsevier/Academic Press on *Cancer Sensitizing Agents for Chemotherapy*, *Sensitizing Agents for Cancer Resistant to Cell Mediated Immunotherapy*, and *Breaking Tolerance to Anti-Cancer Immunotherapy.* Lastly, the Editor is the Editor-in-Chief of the Journal Critical Reviews in Oncogenesis. The Editor has published over 500 research publications and reviews in various scientific journals of high impact.

Acknowledgments: The series editor acknowledges the Department of Microbiology, Immunology, and Molecular Genetics and the UCLA David Geffen School of Medicine for their continuous support. He also acknowledges the assistance of Mr. Rafael Teixeira, acquisitions editor for Elsevier/Academic Press, and the excellent assistance and support of Ms. Samantha Allard, editorial project manager for Elsevier/Academic Press, for their continuous cooperation throughout the development of this book.

Aims and Scope of the Volume

Although great advancements have been acquired in cancer management in recent decades, the cytotoxic drugs such as cisplatin and paclitaxel still play an indispensable role in clinics. Unfortunately, the inevitable chemoresistance severely affects the therapy index and prognosis of patients. The mechanism that leads to chemoresistance is complicated and deeper understanding of this field would pave the way to tackle this issue. Recently, the dysregulation of epigenetic regulation has been confirmed to be involved in this process. Moreover, the epigenetic agents that have shown property of reversing drug resistance in laboratory and clinical trials are one of the promising sensitizing agents.

The main objective of the volume "Epigenetic Regulation in Overcoming Chemoresistance" is to draw a new scenario based on the new scientific and clinical findings about epigenetic regulation in cancer and on the new pharmacological weapons acting on this machinery to overcoming chemoresistance. Furthermore, the volume will give a comprehensive delineation of the interactions between different epigenetic machinery.

The scope of the volume is to provide researchers and oncologists information with undated scientific findings about using epigenetic drugs as chemotherapy sensitizing agents in order to promote the finding of novel epigenetic drug combinations to overcome chemoresistance in cancer.

Dr. Chunfu Wu, PhD
Dr. Lihui Wang, PhD

About the Volume Editors

Dr. Chunfu Wu is a professor at the Department of Pharmacology in Shenyang Pharmaceutical University. After graduation in 1982 he joined Shenyang Pharmaceutical University and was promoted as a full professor in 1994. He received his PhD from Hokkaido Pharmaceutical University in 2000.

The research fields of Professor Wu include the neuropharmacology, pharmacology of traditional Chinese medicine, and tumor pharmacology. For tumor pharmacology, Professor Wu and his group have done great research in identifying epigenetic targets that are involved in chemoresistance or targeted therapy resistance and have discovered epigenetic agents based on these targets to enhance the efficiency of anticancer therapies. Wu's group also has done some investigations in treating cancer using small molecular compounds, such as procaspase-3 activator, HIF inhibitor, ALDH inhibitor, etc.

Based on these investigations, Dr. Wu has published more than 300 research reports in international peer-reviewed journals such as "Molecular Cancer," "Cancer Letters," "Cell Death & Disease," and "Molecular Oncology." He is also the editor in chief of "Asian Journal of Traditional Medicines" and a member on the editorial board of "Acta Pharmacologica Sinica," "Pharmacology, Biochemistry and Behaviors," and "International Journal of Immunopharmacology."

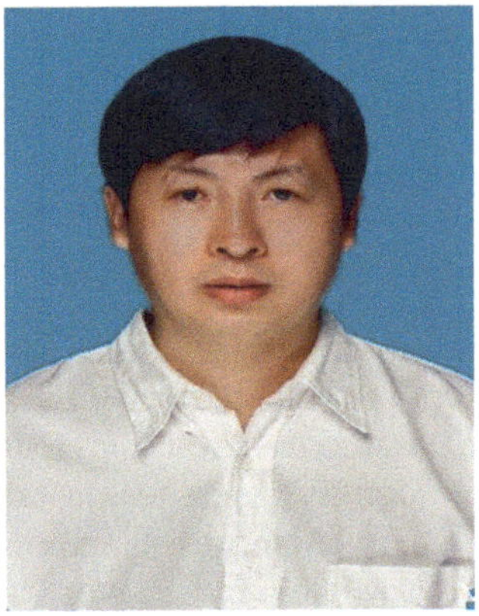

Dr. Lihui Wang is a professor at the Shenyang Pharmaceutical University. He joined Memorial Sloan Kettering Cancer Center in New York as a visiting investigator from the Spring of 2016 to the Summer of 2017.

His research investigations mainly focus on clarifying resistant mechanisms of cancer cells to chemotherapies or targeted therapies as well as identifying small molecule compounds acting on these mechanisms to overcome the anticancer drug resistance. Currently, many of his studies concentrated on discovering epigenetic targets involved in cytotoxic/TKI drug resistance, especially those affecting the histone modification. Another area of research is the development of epigenetic agents based on above targets.

Dr. Wang has published more than 20 reports, as a first or a corresponding author, in international peer-reviewed journals such as "Molecular Cancer," "Clinical Cancer Research," "British Journal of Cancer," and "Cell Death & Disease." He has participated in the editing of the book titled "Traditional Herbal Medicine Research Methods: Identification, Analysis, Bioassay, and Pharmaceutical and Clinical Studies."

Preface

Epigenetic regulation is a reversible process in which the genotype remains unchanged and phenotypes change. It includes DNA methylation, histone modification, noncoding RNA (ncRNA), RNA methylation, etc. Recently, increasing evidences have demonstrated that epigenetic regulation is involved in several aspects of tumor development and progression, especially the chemosensitivity of cancer cells. Given the undesired influence of multidrug resistance, which limits the effect of chemotherapy and leads to poor prognosis, agents targeting the epigenetic modulators become one of the main hotspots of tumor research with the hope to overcoming chemoresistance.

To date, several classes of epigenetic agents such as DNA methyltransferase inhibitors (DNMTi) and histone deacetylase inhibitors (HDACi) have been undergone preclinical and clinical studies in treating hematologic and solid tumors. Delightedly, a few of them have been approved as clinical treatment options for specific tumors and they indeed have shown positive therapy efficiency. However, the mechanisms, applications, and potential of these promising agents are yet to be completely depicted. In this volume titled "A new epigenetic dawn to overcome drug resistance in human cancer," we primarily elucidate the utility of epigenetic agents to enhance the chemotherapy sensitivity.

The volume is divided into eight chapters including the mechanisms and the updated laboratory and clinical trials that use epigenetic agents such as DNMTi, HDACi, HDMi (histone lysine demethylase inhibitors), HMTi (histone lysine methyltransferase inhibitors), and drugs targeting the ncRNA as chemotherapy-sensitizing agents. The experts from the field of cancer chemosensitization also give insights about the role of epigenetic regulation in cancer, the combination therapy involving epigenetic, chemo, and immune drugs as well as the future epigenetic treatment strategies. We truly hope that this delineation of the scientific findings would give readers new inspirations about using the epigenetic weapons to tackle the issue of chemoresistance.

Dr. Chunfu Wu, Ph.D.
Dr. Lihui Wang, Ph.D.
Department of Pharmacology,
Shenyang Pharmaceutical University,
Shenyang, People's Republic of China

Contents

7. The combination strategy with epigenetic therapy, chemotherapy, and immunotherapy

Shijun Wen, Haiwen Wang, and Sixiao Lu

8. Future epigenetic treatment strategies

Lingyue Gao and Lihui Wang

Contributors

Ayumu Asai Artificial Intelligence Research Center, The Institute of Scientific and Industrial Research, Osaka University, Osaka, Japan

Lingyue Gao Department of Pharmacology, Shenyang Pharmaceutical University, Shenyang, People's Republic of China

Takaaki Hirotsu Osaka University Graduate School of Medicine, Center of Medical Innovation and Translational Research, Osaka; Hirotsu Bio Science Inc., Tokyo, Japan

Hideshi Ishii Osaka University Graduate School of Medicine, Center of Medical Innovation and Translational Research, Osaka, Japan

Masamitsu Konno Osaka University Graduate School of Medicine, Center of Medical Innovation and Translational Research, Osaka, Japan

Shini Liu Sun Yat-sen University Cancer Center, State Key Laboratory of Oncology in South China, Collaborative Innovation Center of Cancer Medicine, Guangzhou, Guangdong, People's Republic of China

Sixiao Lu State Key Laboratory of Oncology in South China, Sun Yat-sen University Cancer Center, Guangzhou, People's Republic of China; Division of Biosciences, University College London, London, United Kingdom

Changliang Shan College of Pharmacy, Nankai University, Tianjin, People's Republic of China

Ao Shen Key Laboratory of Molecular Target & Clinical Pharmacology and the State Key Laboratory of Respiratory Disease, School of Pharmaceutical Sciences & The Fifth Affiliated Hospital, Guangzhou Medical University, Guangzhou, People's Republic of China

Jiaqi Song College of Pharmacy, Nankai University, Tianjin, People's Republic of China

Huanran Sun College of Pharmacy, Nankai University, Tianjin, People's Republic of China

Mingming Sun College of Pharmacy, Nankai University, Tianjin, People's Republic of China

Jing Tan Sun Yat-sen University Cancer Center, State Key Laboratory of Oncology in South China, Collaborative Innovation Center of Cancer Medicine, Guangzhou, Guangdong, People's Republic of China; Laboratory of Cancer Epigenome, Division of Medical Sciences, National Cancer Centre Singapore, Singapore; Affiliated Cancer Hospital & Institute of Guangzhou Medical University, Guangzhou, Guangdong, People's Republic of China

Haiwen Wang State Key Laboratory of Oncology in South China, Sun Yat-sen University Cancer Center, Guangzhou, People's Republic of China

Lihui Wang Department of Pharmacology, Shenyang Pharmaceutical University, Shenyang, People's Republic of China

Shijun Wen State Key Laboratory of Oncology in South China, Sun Yat-sen University Cancer Center, Guangzhou, People's Republic of China

Chunfu Wu Department of Pharmacology, Shenyang Pharmaceutical University, Shenyang, People's Republic of China

Cheng Yang Department of Pharmacology, Shenyang Pharmaceutical University, Shenyang, People's Republic of China

Xi-Yong Yu Key Laboratory of Molecular Target & Clinical Pharmacology and the State Key Laboratory of Respiratory Disease, School of Pharmaceutical Sciences & The Fifth Affiliated Hospital, Guangzhou Medical University, Guangzhou, People's Republic of China

Jiayu Zhang Department of Pharmacology, Shenyang Pharmaceutical University, Shenyang, People's Republic of China

CHAPTER

1

The role of epigenetic regulation in cancer

Jiayu Zhang, Cheng Yang, and Chunfu Wu

Department of Pharmacology, Shenyang Pharmaceutical University, Shenyang, People's Republic of China

Abstract

Cancer has always been one of the main threats to human health. Looking for its potential pathogenic mechanism is a major challenge facing humanity. Epigenetic mechanisms are essential to maintain specific gene expression patterns and the normal development of life individuals. They mainly regulate gene function and expression levels through DNA methylation, histone modification, and RNA regulation. When the epigenetic regulation is abnormal, it interferes with the expression and function of normal genes to a certain extent, which may lead to the occurrence of tumors. Therefore, epigenetic modification is closely related to tumorigenesis. In order to understand the nature of cancer and open up novel treatment methods, it is particularly important to figure out the mechanism of epigenetics in cancer. This article outlines the research content of epigenetics, focusing on its relationship with tumorigenesis. At the same time, the application of epigenetics in cancer therapy is prospected.

Abbreviations

5mC	5-methylcytosine
AML	acute myeloid leukemia
BASP1	brain acid-soluble protein 1
CH3	methyl
CircRNA	circular RNA
CpG	cytosine-phosphate-guanine
DNMT	DNA methyltransferase
H3K4	lysine 4 of histone H3
H3K9	lysine 9 of histone H3
H3K27	lysine 27 of histone H3
H3K36	lysine 36 of histone H3
H3K79	lysine 79 of histone H3
H4K20	lysine 20 of histone H4
HAT	histone acetyltransferase

Epigenetic Regulation in Overcoming Chemoresistance
https://doi.org/10.1016/B978-0-12-823684-0.00005-0

HDAC	histone deacetylase
HDM	histone demethyltransferase
hMLH1	human mutL homolog-1
HOTAIR	HOX transcript antisense RNA
KDM	lysine demethylase
Kme1	lysine mono-methylation
Kme2	lysine di-methylation
Kme3	lysine tri-methylation
KMT	lysine methyltransferase
LncRNA	long noncoding RNA
m6A	N6-methyl adenine
miRNA	microRNA
ncRNA	noncoding RNA
PRMT	protein arginine methyltransferase
Rme1/MMA	N-monomethylarginine
Rme2	di-methylarginine
Rme2a/ADMA	asymmetric dimethylarginine
Rme2s/SDMA	symmetric dimethylarginine
SAM	S-adenosyl-L-methionine

Conflict of interest

No potential conflicts of interest were disclosed.

Introduction

Tumor is currently one of the most important diseases that harms humans. Previously, scientists only roughly regarded it as a type of genetic disease related to gene mutations, especially the activation of oncogenes and the inactivation of tumor suppressor genes [1]. However, with the in-depth study of cancer, it is found that epigenetic changes in all types of cancer far exceed somatic gene mutations [2]. Genome sequencing and other methods have shown that many proteins that code for regulating chromatin or DNA have changed in cancer. People have a more precise understanding of the complexity of chromatin function and its role in tumorigenesis, and define the process involved in changing gene expression and inheritance without alteration of DNA sequence as epigenetics [3]. Since then, epigenetics has been a major research focus in gene expression regulation. Moreover, abnormal epigenetic regulation does play a decisive role in tumorigenesis, as well as invasion and migration. This change can not only be used as a marker for early tumor diagnosis, but may also become a new target for tumor prevention and treatment. This chapter mainly describes the role of epigenetics in tumors.

The research of epigenetics mainly focuses on the regulation of DNA and histones—the two major macromolecules that together constitute chromatin. According to the location and type of modification, it can be roughly divided into DNA methylation, histone modification, and RNA regulation [4].

DNA methylation

DNA methylation is the most extensive and in-depth study of important physiological reactions in epigenetic modification, which is catalyzed by DNA methyltransferase (DNMT), adding the methyl (CH3) group to the C5 position in the cytosine-phosphate-guanine (CpG) dinucleotide pair to make the cytosine ring form 5-methylcytosine (5mC), which could lead to changes in gene expression, but the DNA sequence and gene products do not change [5].

There are five types of DNMTs which catalyze different methylation reactions, namely, DNMT1, RNA methyltransferase DNMT2, de novo methylases DNMT3A, and DNMT3B, which maintain the methylation state after DNA replication, and one that lacks enzymes, the catalytically active domain of DNMT3 analogous protein DNMT3L [6]. In mammals, the promoter of active genes is usually demethylated, while the promoter of silent genes is hypermethylated. Studies have shown that DNA methylation is involved in a variety of biological processes, such as X chromosome inactivation, transcription factor silencing, heterochromatin formation, genetic imprinting, genome stability, and embryonic development [7]. The alteration of DNA methylation status occurs frequently in cancer. Whole-genome hypomethylation and promoter hypermethylation are the hallmarks of many cancers that cause tumors. DNA hypomethylation promotes chromosomal instability and gene overexpression. The imbalance of DNA methylation status in tumors may cause genetic changes in gene expression and form the molecular basis of genome instability [8]. DNA hypermethylation can cause changes in chromosome structure, which cannot be combined with transcription factors and changes in gene expression profiles, including the silencing of tumor suppressor genes, endogenous retroviruses and tumor antigens, and the upregulation of oncogene expression [9, 10].

Hypomethylation is common in some solid tumors such as digestive system tumors and hematological system tumors, and the higher the malignant degree, the more obvious it is. It is found that there is a negative correlation with gene expression in BCL-2, an antiapoptosis gene of B-cell chronic lymphocytic leukemia, and *K-RAS*, a proto-oncogene of lung cancer and colon cancer [11, 12]. However, hypermethylation is related to drug resistance, tumor cell proliferation and angiogenesis, and other malignant phenotypes. Increased expression of DNMT3B in ovarian cancer cells will activate the mismatch repair gene human mutL homolog-1 (hMLH1) and p53 and other tumor suppressor genes. The increase in methylation level of the subregion leads to the decrease of the abovementioned gene expression and promotes the development of tumors [13]. In acute myeloid leukemia (AML), DNMT3a recruits the Brain acid-soluble protein 1 (BASP1) promoter sequence, which triggers DNA methylation-mediated BASP1 silencing, while the downregulation of BASP1 enables the proliferation and colony forming ability of AML cells enhanced [14]. Thus, DNA methylation is essential for life activities.

Histone modification

Different from DNA modification, histone modification can be divided into histone methylation and demethylation, and histone acetylation and deacetylation. According to different modified sites, there are some differences in functions [15]. Histone methylation can be divided into lysine methylation and arginine methylation according to different amino acids. According to the amount of methyl groups (S-adenosyl-L-methionine, SAM) added, mono-, di-, and trimethylated derivatives (Kme1, Kme2, Kme3) were formed in lysine sites, while mono- and dimethyl derivatives (Rme1/MMA, Rme2) were formed in arginine sites. However, unlike lysine, due to the spatial structure of arginine, dimethyl derivatives produce two different products: asymmetric dimethylarginine (Rme2a or ADMA) and symmetric dimethylarginine (Rme2s or SDMA). This reaction only slightly changes the primary structure of the protein but greatly increases the hydrophobicity and molecular information of the protein, thus changing the binding and modification ability of the protein, and finally affecting the downstream gene expression, thus achieving the regulation of genetic information and further affecting the biological functions of organisms [16].

Lysine methylation is catalyzed by protein lysine methyltransferase (KMT) and removed by protein lysine demethylase (KDM). In humans, the main methylation occurs in lysine 4 (H3K4), lysine 9 (H3K9), lysine 27 (H3K27), lysine 36 (H3K36), and lysine 79 (H3K79) of histone H3 and lysine 20 (H4K20) of histone H4. Some of these sites are generally considered to be involved in regulating the transcriptional activation of downstream genes, such as H3K4; some are considered to be related to regulating the transcriptional repression of downstream genes, such as H3K27. Moreover, the methylation of the same site may be completed by many different KMTs. SETD2 synthesizes H3K36me3 in humans and participates in tumor suppression. While NSD1, NSD2, NSD3, and ASH1L jointly regulate H3K36me2, it can also be completed by the same KMTs; for example, DOT1L catalyzes H3K79 me/me2/me3, but the current research on lysine methyltransferase and cancer mainly focuses on several KMTs; for example, EZH2 catalyzes H3K27me3, which inhibits the transcription of downstream E-cadherin, P16 INK4α, P57, PSP94, and other tumor suppressor genes, resulting in cancer [17]. G9a catalyzes the effect of H3K9me/me2 on the expression of RARRES3, HP1α, APC2, and other genes to promote cancer [18, 19]. SETD2, the occurrence of inactivating mutations that lead to cancer, and DOT1L participate in multiple processes in breast cancer [20, 21], ovarian cancer, and other nausea cancers [22]. They are overexpressed in the body, inactivating mutations or gain-of-function mutations, which produce abnormal regulation on downstream genes in different forms and eventually lead to cancer [23].

Arginine methylation is also catalyzed by protein arginine methyltransferase (PRMT) and removed by protein arginine demethylase. However, unlike a large number of KMTs with different kinds, PRMTs has only nine family members, namely PRMT1–9. These 9 family members can all catalyze the mono methylation of arginine, which can then be further classified according to the type of dimethylarginine catalyzed: Type I PRMTs (PRMT1–4, 6, 8) produce asymmetric dimethylarginine. Type II PRMTs (PRMT5, 9) catalyze the production of symmetric dimethylarginine [24]. PRMT7 is special and only has the ability to catalyze mono methylation. PRMT1 is the most active member of the family, accounting for 85% of the activity of all PRMT members [25], whereas PRMT5 is the most important type II PRMT in this family [26]. Compared with KMTs, the research on tumor of PRMTs is limited. However, PRMT5 has been proved to promote the occurrence and development of cancer by participating in cell survival and differentiation, and other family members have also been proved to participate in tumorigenesis by catalyzing the methylation of nonhistone proteins [27–29].

It can be seen from the above that the roles of KMTs and PRMTs in cancer are diverse and complex, and they have different and even seemingly contradictory roles according to different sites and downstream target genes. Moreover, methylation and demethylation in the human body require a precise balance [30]. The loss of control of KMTs and PRMTs will undoubtedly cause the loss of control of histone demethyltransferase (HDM). Therefore, such characteristics promote HDM to participate in tumorigenesis. In different cancers, we can observe the dysregulation of certain lysine demethylases. For example, LSD1 is not highly expressed in some breast cancers, but it is highly expressed in many other solid tumors and leukemias [31–33]. In a single cancer type, we can also observe the dysregulation of different demethylases. For example, the expression of KDM5C, KDM4A, or KDM4B is upregulated in breast cancer [34]. The above research shows that demethylases have the same complex regulatory mechanism in different cancers. Therefore, maintaining a good balance of methylation levels in the body by targeting methyltransferase or demethyltransferase is an effective cancer treatment [35].

In addition to histone methylation, acetylation still plays an important role in histone modification. However, compared to the complex and diverse functions of methylation, acetylation is relatively simple. Histone acetylation is regulated by histone acetyltransferase (HAT) and histone deacetylase (HDAC). HAT adds acetyl groups to the lysine residues of histones to neutralize the positive charge and relax the nucleosomes, thereby promoting the transcriptional activity of downstream genes [36]. Although the specific regulation mechanism is not clear, it can be confirmed that it is related to cancer [37, 38]. HDAC, which has the opposite function, is more closely related to cancer. The HDAC family has 11 members and can be roughly divided into four categories: I (HDAC1–3,8), IIa (HDAC4,5,7,9), IIb (HDAC6,10), and IV (HDAC11). They are roughly similar in function. After deacetylation, it is easier to transcriptional inhibition by enhancing the condensed state of chromatin [39]. But there is a slight difference in other aspects. For example, HDAC I is a ubiquitously expressed HDAC that is located in the nucleus. IIa is a type of HDAC with relatively restricted expression. Among them, HDAC5 and 9 are expressed in the muscle and heart, HDAC4 is expressed in the bone, and HDAC7 is expressed in endothelial cells and thymocytes. Little is known about HDAC10 in IIb, but it can be determined that HDAC6 is the most important cytoplasmic deacetylase [40]. HDAC11 belongs to class IV HDAC and is enriched in the brain, heart, and muscle, but its function is still unclear [41]. The current research on HDAC and cancer is mainly focused on HDAC1–3,8 and HDAC6, and its inhibitors have been approved by the FDA for treating cancer.

RNA regulation

In addition to the methylation of genetic material DNA, there is another genetic material RNA that can be methylated, which accounts for more than 60% of all RNA modifications. N6-methyl adenine (m6A) is the most common posttranscriptional modification in eukaryotic cells, mainly concentrated in the promoter region, stop codon region, and RRACH motif of mRNA. Studies have shown that m6A modification participates in regulating many vital biological processes, such as stem cell self-renewal, tissue differentiation, DNA damage response, etc. Therefore, when the metabolic process regulated by RNA is abnormal, many diseases, such as cancer, will occur immediately [42]. In addition, m6A methylation provides more possibilities for early diagnosis and oncotherapy, and it is also a relatively new research direction in recent years [43].

Another important way is noncoding RNA (ncRNA) regulation, which refers to functional RNA that can be transcribed but cannot be translated into protein, such as long noncoding RNA (lncRNA), microRNA (miRNA), circular RNA (circRNA), etc. [44]. LncRNA is considered as a noncoding RNA with a length of more than 200 nucleotides produced by RNA polymerase II. The regulation mode of lncRNA is diversified, which can combine with DNA, RNA, or protein and regulate gene expression in many different modes of action. The regulation mode may be cis or trans, which plays a role in chromatin modification, transcription activation, and translation regulation and many other life processes, such as chromatin remodeling mediated by HOX transcript antisense RNA (HOTAIR) [45].

In addition, miRNA can not only inhibit the expression of target gene or induce its degradation and translation inhibition at the posttranscriptional level, but also form complementary sequences with the promoter of the target gene at the transcriptional level, which leads to gene silencing [46]. Therefore, miRNA can induce double inhibition of target mRNA. In addition, circRNA is a single-stranded circ RNA molecule formed by reverse splicing of exons in pre-mRNA. Some circRNAs can regulate important tumor characteristics, including tumor cells. Proliferation, apoptosis, metastasis, metabolism, aging, and drug resistance are even related to the prognosis of cancer patients [47].

Conclusion

No matter which epigenetic modifications are mentioned above, when they are abnormal, they may lead to the occurrence of cancer, so more and more researchers focus on this field, hoping to study the pathogenesis of cancer and the best way to treat cancer through epigenetics, which is rapidly changing the traditional methods of treating cancer. In addition, we found that both gene mutation and epigenetic abnormality can lead to cancer, but unlike gene mutation, epigenetic changes are reversible in nature and can cause gene reexpression through treatment. At the same time, aiming at the traditional carcinogenic route and epigenetic treatment is a good way for most tumors, which is undoubtedly an exciting way for cancer targeted therapy [48].

References

[1] Sato M, Shames DS, Gazdar AF, Minna JD. A translational view of the molecular pathogenesis of lung cancer. J Thorac Oncol 2007;2(4):327–43.

[2] Hoadley KA, Yau C, Wolf DM, Cherniack AD, Tamborero D, Ng S, et al. Multiplatform analysis of 12 cancer types reveals molecular classification within and across tissues of origin. Cell 2014;158(4):929–44.

[3] Han M, Jia L, Lv W, Wang L, Cui W. Epigenetic enzyme mutations: role in tumorigenesis and molecular inhibitors. Front Oncol 2019;9:194.

[4] Vicente-Duenas C, Hauer J, Cobaleda C, Borkhardt A, Sanchez-Garcia I. Epigenetic priming in cancer initiation. Trends Cancer 2018;4(6):408–17.

[5] Zhang J, Yang C, Wu C, Cui W, Wang L. DNA methyltransferases in cancer: biology, paradox, aberrations, and targeted therapy. Cancers (Basel) 2020;12(8).

[6] Cheng X, Blumenthal RM. Mammalian DNA methyltransferases: a structural perspective. Structure 2008;16(3):341–50.

[7] Dor Y, Cedar H. Principles of DNA methylation and their implications for biology and medicine. Lancet 2018;392(10149):777–86.

[8] Feinberg AP. Phenotypic plasticity and the epigenetics of human disease. Nature 2007;447(7143):433–40.

[9] Chen RZ, Pettersson U, Beard C, Jackson-Grusby L, Jaenisch R. DNA hypomethylation leads to elevated mutation rates. Nature 1998;395(6697):89–93.

[10] Baer C, Claus R, Plass C. Genome-wide epigenetic regulation of miRNAs in cancer. Cancer Res 2013;73(2):473–7.

[11] Tellez CS, Grimes MJ, Picchi MA, Liu Y, March TH, Reed MD, et al. SGI-110 and entinostat therapy reduces lung tumor burden and reprograms the epigenome. Int J Cancer 2014;135(9):2223–31.

[12] Ruiz-Magana MJ, Rodriguez-Vargas JM, Morales JC, Saldivia MA, Schulze-Osthoff K, Ruiz-Ruiz C. The DNA methyltransferase inhibitors zebularine and decitabine induce mitochondria-mediated apoptosis and DNA damage in p53 mutant leukemic T cells. Int J Cancer 2012;130(5):1195–207.

[13] Zhao L, Shou H, Chen L, Gao W, Fang C, Zhang P. Effects of ginsenoside Rg3 on epigenetic modification in ovarian cancer cells. Oncol Rep 2019;41(6):3209–18.
[14] Zhou L, Fu L, Lv N, Liu J, Li Y, Chen X, et al. Methylation-associated silencing of BASP1 contributes to leukemogenesis in t(8;21) acute myeloid leukemia. Exp Mol Med 2018;50(4):44.
[15] Sun W, Lv S, Li H, Cui W, Wang L. Enhancing the anticancer efficacy of immunotherapy through combination with histone modification inhibitors. Genes (Basel) 2018;9(12).
[16] Yang C, Zhang J, Ma Y, Wu C, Cui W, Wang L. Histone methyltransferase and drug resistance in cancers. J Exp Clin Cancer Res 2020;39(1):173.
[17] Cao R, Wang L, Wang H, Xia L, Erdjument-Bromage H, Tempst P, et al. Role of histone H3 lysine 27 methylation in Polycomb-group silencing. Science (New York, NY) 2002;298(5595):1039–43.
[18] Wei L, Chiu DK, Tsang FH, Law CT, Cheng CL, Au SL, et al. Histone methyltransferase G9a promotes liver cancer development by epigenetic silencing of tumor suppressor gene RARRES3. J Hepatol 2017;67(4):758–69.
[19] Zhang K, Wang J, Yang L, Yuan Y-C, Tong TR, Wu J, et al. Targeting histone methyltransferase G9a inhibits growth and Wnt signaling pathway by epigenetically regulating HP1α and APC2 gene expression in non-small cell lung cancer. Mol Cancer 2018;17(1):153.
[20] Zhu X, He F, Zeng H, Ling S, Chen A, Wang Y, et al. Identification of functional cooperative mutations of SETD2 in human acute leukemia. Nat Genet 2014;46(3):287–93.
[21] Zhang YL, Sun JW, Xie YY, Zhou Y, Liu P, Song JC, et al. Setd2 deficiency impairs hematopoietic stem cell self-renewal and causes malignant transformation. Cell Res 2018;28(4):476–90.
[22] Liu D, Zhang XX, Li MC, Cao CH, Wan DY, Xi BX, et al. C/EBPbeta enhances platinum resistance of ovarian cancer cells by reprogramming H3K79 methylation. Nat Commun 2018;9(1):1739.
[23] Nassa G, Salvati A, Tarallo R, Gigantino V, Alexandrova E, Memoli D, et al. Inhibition of histone methyltransferase DOT1L silences ERα gene and blocks proliferation of antiestrogen-resistant breast cancer cells. Sci Adv 2019;5(2):eaav5590.
[24] Stopa N, Krebs JE, Shechter D. The PRMT5 arginine methyltransferase: many roles in development, cancer and beyond. Cell Mol Life Sci 2015;72(11):2041–59.
[25] Yamaguchi A, Kitajo K. The effect of PRMT1-mediated arginine methylation on the subcellular localization, stress granules, and detergent-insoluble aggregates of FUS/TLS. PLoS One 2012;7(11), e49267.
[26] Musiani D, Bok J, Massignani E, Wu L, Tabaglio T, Ippolito MR, et al. Proteomics profiling of arginine methylation defines PRMT5 substrate specificity. Sci Signal 2019;12(575).
[27] Avasarala S, Van Scoyk M, Karuppusamy Rathinam MK, Zerayesus S, Zhao X, Zhang W, et al. PRMT1 is a novel regulator of epithelial-mesenchymal-transition in non-small cell lung cancer. J Biol Chem 2015;290(21):13479–89.
[28] Hsu MC, Tsai YL, Lin CH, Pan MR, Shan YS, Cheng TY, et al. Protein arginine methyltransferase 3-induced metabolic reprogramming is a vulnerable target of pancreatic cancer. J Hematol Oncol 2019;12(1):79.
[29] Wong TL, Ng KY, Tan KV, Chan LH, Zhou L, Che N, et al. CRAF methylation by PRMT6 regulates aerobic glycolysis-driven hepatocarcinogenesis via ERK-dependent PKM2 nuclear relocalization and activation. Hepatology 2020;71(4):1279–96.
[30] Jambhekar A, Anastas JN, Shi Y. Histone lysine demethylase inhibitors. Cold Spring Harb Perspect Med 2017;7(1).
[31] Wang Y, Zhang H, Chen Y, Sun Y, Yang F, Yu W, et al. LSD1 is a subunit of the NuRD complex and targets the metastasis programs in breast cancer. Cell 2009;138(4):660–72.
[32] Hayami S, Kelly JD, Cho HS, Yoshimatsu M, Unoki M, Tsunoda T, et al. Overexpression of LSD1 contributes to human carcinogenesis through chromatin regulation in various cancers. Int J Cancer 2011;128(3):574–86.
[33] Schildhaus HU, Riegel R, Hartmann W, Steiner S, Wardelmann E, Merkelbach-Bruse S, et al. Lysine-specific demethylase 1 is highly expressed in solitary fibrous tumors, synovial sarcomas, rhabdomyosarcomas, desmoplastic small round cell tumors, and malignant peripheral nerve sheath tumors. Hum Pathol 2011;42(11):1667–75.
[34] Wang Q, Wei J, Su P, Gao P. Histone demethylase JARID1C promotes breast cancer metastasis cells via down regulating BRMS1 expression. Biochem Biophys Res Commun 2015;464(2):659–66.
[35] Berry WL, Shin S, Lightfoot SA, Janknecht R. Oncogenic features of the JMJD2A histone demethylase in breast cancer. Int J Oncol 2012;41(5):1701–6.
[36] Dekker FJ, Haisma HJ. Histone acetyl transferases as emerging drug targets. Drug Discov Today 2009;14(19–20):942–8.
[37] Devaiah BN, Case-Borden C, Gegonne A, Hsu CH, Chen Q, Meerzaman D, et al. BRD4 is a histone acetyltransferase that evicts nucleosomes from chromatin. Nat Struct Mol Biol 2016;23(6):540–8.

[38] Semer M, Bidon B, Larnicol A, Caliskan G, Catez P, Egly JM, et al. DNA repair complex licenses acetylation of H2A.Z.1 by KAT2A during transcription. Nat Chem Biol 2019;15(10):992–1000.
[39] Richon VM. Cancer biology: mechanism of antitumour action of vorinostat (suberoylanilide hydroxamic acid), a novel histone deacetylase inhibitor. Br J Cancer 2006;95(S1):S2–6.
[40] Zhang Y, Kwon S, Yamaguchi T, Cubizolles F, Rousseaux S, Kneissel M, et al. Mice lacking histone deacetylase 6 have hyperacetylated tubulin but are viable and develop normally. Mol Cell Biol 2008;28(5):1688–701.
[41] Gao L, Cueto MA, Asselbergs F, Atadja P. Cloning and functional characterization of HDAC11, a novel member of the human histone deacetylase family. J Biol Chem 2002;277(28):25748–55.
[42] Yue Y, Liu J, He C. RNA N6-methyladenosine methylation in post-transcriptional gene expression regulation. Genes Dev 2015;29(13):1343–55.
[43] Sun T, Wu R, Ming L. The role of m6A RNA methylation in cancer. Biomed Pharmacother 2019;112:108613.
[44] Kaikkonen MU, Adelman K. Emerging roles of non-coding RNA transcription. Trends Biochem Sci 2018;43(9):654–67.
[45] Quinn JJ, Chang HY. Unique features of long non-coding RNA biogenesis and function. Nat Rev Genet 2016;17(1):47–62.
[46] Pu M, Li C, Qi X, Chen J, Wang Y, Gao L, et al. MiR-1254 suppresses HO-1 expression through seed region-dependent silencing and non-seed interaction with TFAP2A transcript to attenuate NSCLC growth. PLoS Genet 2017;13(7), e1006896.
[47] Zhang M, Huang N, Yang X, Luo J, Yan S, Xiao F, et al. A novel protein encoded by the circular form of the SHPRH gene suppresses glioma tumorigenesis. Oncogene 2018;37(13):1805–14.
[48] Song S-H, Han S-W, Bang Y-J. Epigenetic-based therapies in cancer: progress to date. Drugs 2011;71(18): 2391–403.

CHAPTER

2

DNA methyltransferase inhibitors (DNMTis) as sensitizing agents to overcome chemoresistance

Shini Liu[a] *and Jing Tan*[a,b,c]

[a]Sun Yat-sen University Cancer Center, State Key Laboratory of Oncology in South China, Collaborative Innovation Center of Cancer Medicine, Guangzhou, Guangdong, People's Republic of China [b]Laboratory of Cancer Epigenome, Division of Medical Sciences, National Cancer Centre Singapore, Singapore [c]Affiliated Cancer Hospital & Institute of Guangzhou Medical University, Guangzhou, Guangdong, People's Republic of China

Abstract

DNA methylation is one of the most characterized epigenetic modifications, which plays a vital role in regulating gene transcription, chromatin remodeling, and genome stability. Aberrant DNA methylation promotes cancer progression and drug resistance. Inhibitors targeting DNA methyltransferases (DNMTs), the key catalytic enzymes in DNA methylation, have evidently shown therapeutic efficacy in many malignancies. Chemotherapy is one of the major treatment regimens for cancer patients. However, acquired drug resistance invariably hinders the clinical effect. Increasing evidence has shown that aberrant DNA methylation is involved in chemoresistance and DNMTis have the promising potential to reverse the chemoresistance. Here we summarize recent advances in DNMTis as novel sensitizers to overcome the resistance to chemotherapy.

Abbreviations

AML	acute myeloid leukemia
ATRA	all-trans retinoic acid
AZA	azacitidine
Bp	base pair
CHOP	cyclophosphamide + doxorubicin hydrochloride + vincristine + prednisone
CpG	cytosine–phosphate–guanine
CSC	cancer stem cells
DAC	decitabine
DNA	deoxyribonucleic acid

https://doi.org/10.1016/B978-0-12-823684-0.00001-3

DNMT	DNA methyltransferase
DNMTis	DNA methyltransferase inhibitors
MGMT	6-methylguanine-DNA methyltransferase
RCC	renal cell carcinoma
RNA	ribonucleic acid

Conflict of interest

No potential conflicts of interest were disclosed.

Aberrant DNA methylation in cancer

Epigenetics is the process that reversible and heritable patterns of gene expression occur without any change of DNA sequence [1], which mainly include DNA methylation, histone modification, chromatin remodeling, and RNA interference [2]. DNA methylation is one of the most common epigenetic modifications, arising from the addition of a methyl group from S-adenosylmethionine (acts as a methyl donor) to the fifth carbon of the cytosine DNA base [3]. Aberrant DNA methylation is frequently found in human malignancies, which can be manifested as extensive genome-wide DNA hypomethylation or DNA hypermethylation at specific gene loci. Widespread DNA hypomethylation mainly occurs in DNA repetitive sequence and gene coding regions, leading to genome instability and aberrant activation of oncogenes and transposable elements. In contrast, DNA hypermethylation generally occurs in cytosine-phosphate-guanine (CpG) island, resulting in transcription depression including tumor suppressor genes (TSGs) and tumor-associated antigen genes. CpG islands are defined as >200 bp long region of DNA with a GC ratio of more than 50% and they are often found at promoters. Hypermethylation in the CpG island hinders the accessibility of transcription factors to promoters and promotes methyl-CpG binding domain (MBD) to bind the methylated DNA, leading to recruitment of additional corepressors and thereby leading to gene silencing [4]. Notably, aberrant losses and gains of DNA methylation in regulatory regions of the genome, such as gene enhancers, can also influence gene expression and cancer development [5, 6].

Deregulation of DNA methyltransferases in cancer

DNA methylation is mediated by a conserved family of DNA methyltransferases including DNMT1, DNMT2, DNMT3a/b, and DNMT3L in mammalian cells [7]. DNMT1 recognizes hemimethylated DNA, functioning as a maintenance DNA methyltransferase that is required for stable inheritance of methylation patterns, while DNMT3a and DNMT3b are essential for de novo methylation. DNMT2 and DNMT3L are noncatalytic DNMTs. The former is involved in catalyzing tRNA methylation [8] and its ability to methylate DNA is controversial, while DNMT3L participates in DNA methylation in a noncatalytic manner [9].

Dysregulation of DNMTs has been found in various cancers. Overexpression of DNMT1, DNMT3a, and DNMT3b has been found to be associated with poor prognosis in various cancers [10], such as acute myeloid leukemia (AML), glioma, cervical cancer, and hepatocellular carcinomas. Additionally, frequent DNMT3a mutations were found in various hematological malignancies, such as AML (28%) [11–13] and angioimmunoblastic T-cell lymphoma (AITL)

(33%) [14]. The frequent activating mutation is involved in the development of hematological malignancies since its key role is in regulating the differentiation of hematopoietic stem cells [15–18].

DNA methyltransferase inhibitors (DNMTis) in cancer therapy

Based on the important role of DNA methyltransferase in aberrant DNA methylation and the reversible nature of DNA methylation, targeting DNMTs has attracted notable attention for cancer treatment. DNMTis are classified into two categories: nucleoside analogues and nonnucleoside inhibitors. The main approved and underinvestigated DNMTis are listed in Table 1. Nucleoside analogues are derivatives of cytidine which can be turned into

TABLE 1 The main approved and under-investigated DNMTis.

DNMTi	Chemical characteristics	Mode of action	Chemical structure
Decitabine	Ribonucleoside analogue	A nucleic acid synthesis inhibitor and hypomethylates DNA not RNA by inhibiting DNA methyltransferase	
Azacitidine	Ribonucleoside analogue	A chemical analog of cytidine that inhibits DNA methyltransferase at a lower dosage but incorporates RNA and DNA at a higher dosage	
Thioguanine	Nonnucleoside analogue	A purine antimetabolite inhibits DNMT1 activity through ubiquitin-targeted degradation. Its incorporation alters the DNA cleavage induced by topoisomerase II in the presence and the absence of etoposide	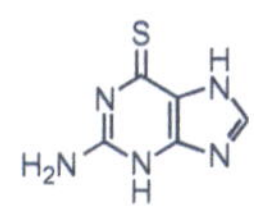
CM272	Nonnucleoside analogue	A novel first-in-class dual reversible inhibitor of G9a (GLP) and DNMTs	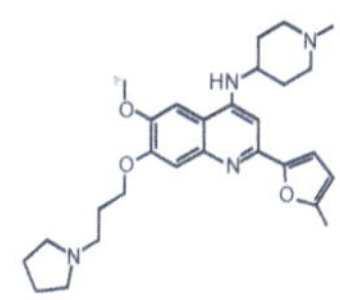
Bobcat339 hydrochloride	Nonnucleoside analogue	A selective cytosine-based TET enzyme inhibitor which reduce DNA 5-hydroxymethylcytosine abundance by inhibiting TET enzyme function	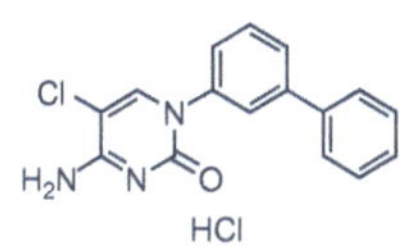
SGI-1027	Nonnucleoside analogue	Inhibits DNA methylation by directly inhibiting DNMTs, and results in selective degradation of DNMT1	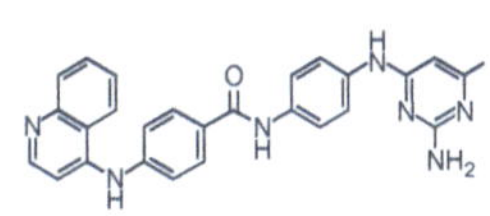
RG108	Nonnucleoside analogue	Acts by direct binding to the methyltransferase enzyme whereby it blocks the enzyme active site	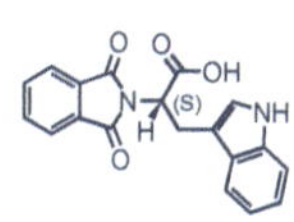

Continued

TABLE 1 The main approved and under-investigated DNMTis—cont'd

DNMTi	Chemical characteristics	Mode of action	Chemical structure
(−)-Epigallocatechin Gallate	Nonnucleoside analogue	The main catechin extraction of green tea that inhibits telomerase and DNA methyltransferase	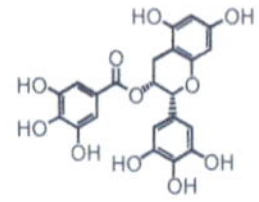
Guadecitabine (SGI-110)	Ribonucleoside analogue	A next-generation hypomethylating agent whose active metabolite decitabine has a longer in-vivo exposure time than intravenous decitabine	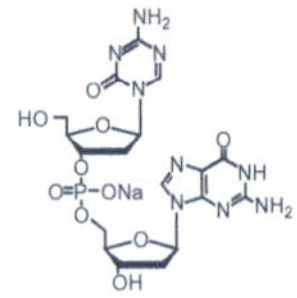
MG98	Nonnucleoside analogue	An antisense oligonucleotide to DNA methyltransferase 1 (DNMT1), which has been shown to reduce CpG island methylation	N/A

nucleotides and incorporated into newly synthesized DNA or RNA, causing enzyme trapping and degradation through the formation of a covalent complex with DNMTs [19]. Azacitidine (AZA) and decitabine (DAC) are the most promising DNMTis that have been approved by the Food and Drug Administration (FDA) and European Medicines Agency (EMA) for the treatment of various hematological malignancies [20]. In addition, multiple preclinical studies have shown their significant antitumor effect in solid tumors [21]. For instance, Tsai et al. showed that nanomolar doses of decitabine and azacitidine exert antitumor effects on epithelial tumor cells through genome-wide changes in DNA methylation and gene expression that inhibit multiple oncogenic signaling pathways [22]. However, adverse side effects of DNMTi usage deserve particular attention. AZA has been proven to cause fetal abnormalities and decreased male fertility, especially at high doses, and DAC is found to induce myelosuppression at all doses. To circumvent this toxicity, nonnucleoside DNMTis which target the catalytic domain only to compete for DNA and/or SAM are being developed. As a relatively less cytotoxic alternative, some of them have exhibited robust anticancer effect. For instance, (−)-epigallocatechin-3-gallate (EGCG) has been shown to reactivate multiple key tumor suppressor genes, including hMLH1, P16, and RA, in various human cancers [23]. Another nonnucleoside demethylating agent SGI-1027 was reported to not only degrade DNMT1 in colon cancer RKO cells, but also result in the demethylation of the CDKN2A gene promoter, reactivating suppressed genes [24, 25]. However, most of the nonnucleoside inhibitors have relatively low bioactivity at micromolar levels and/or fail to achieve sufficient inhibitory effect against DNMTs [26]. Thus, recent studies have shown emerging interest in the development of selective DNMTis with better specificity and minimal off-target effects. MG98 is a phosphorothioate antisense oligodeoxynucleotide which is able to prevent DNMT1 mRNA translation. It was reported to be more effective than 5-AzaCdR in DNA demethylation inhibition [27]. But these controversies exist on whether inhibiting either DNMT alone is sufficient to induce hypomethylation. Cooperation between different DNA methyltransferases implies the need to inhibit several of them simultaneously for optimal clinical benefit [28]. In summary, although many effective DNMTis have been developed, further investigation needs to be conducted to discover new types of DNMTis to overcome current limitations.

DNMTi as a single drug still needs further exploration, but numerous studies have shown that DNMTi combined with other reagents could maximize DNMTi clinical benefit, especially chemotherapy. This chapter will reveal and explore how current DNMTis are used as sensitizing agents for cancer chemotherapy and hopefully inspire further investigation and clinical research.

DNMTi as sensitizing agents for chemotherapy

Cytotoxic therapy is basically the first-line therapeutic strategy for cancer treatment merely secondary to surgical resection in various cancers [29]. However, the therapeutic efficiency is negatively influenced by the inevitable development of resistance. The underlying mechanisms are complex and result from cell extrinsic factors (tumor microenvironment) and intrinsic factors such as enhanced antiapoptosis capability, abnormal DNA damage repair abilities, cancer stem cells, and changes in drug metabolism, detoxification of therapeutic agents (drug-target interactions), which cause a series of abnormalities in gene expression regulation. The contribution of epigenetic modifications, especially abnormal DNA methylation, to resistance has drawn wide attention in recent years. Global and regional aberrant DNA methylation both promote chemoresistance. Thus, DNMTis have been used as potential chemo-sensitizers in various cancers and clinical tests. The representative ongoing clinical trials of DNMTis combined with chemotherapy are listed in Table 2.

Global aberrant DNA methylation induces chemoresistance

Various studies have shown that global aberrant DNA methylation causes tumor cells to gain resistance to chemotherapy that can be overcome by using DNMTis. For instance, Segura-Pacheco et al. detected a significant increase in the level of whole genome methylation in adriamycin-resistant breast cancer cells and the resistance can be partially restored by using a demethylating agent hydralazine [30]. Azacitidine showed synergistic effects with both docetaxel and cisplatin in aggressive prostate cancer models by upmodulating a series of gene expression, such as p16INKA, Bax, Bak, p21/WAF1, and p27/KIP1, and inhibiting the activation of Akt activity and the expression of cyclin D1, Bcl-2, and Bcl-XL [31]. Another DNA methyltransferase inhibitor SGI-110 was also determined to resensitize platinum-resistant ovarian cancer cells to cisplatin (CDDP) and induced both global (LINE1) and gene-specific hypomethylation, including TSGs (RASSF1A), proposed drivers of ovarian cancer cisplatin resistance (MLH1 and ZIC1), differentiation-associated genes (HOXA10 and HOXA11), and transcription factors (STAT5B) [32].

Responses of cancer cells to drugs are not exclusively determined by their intrinsic characteristics but are also regulated by deriving signals from the tumor microenvironment. Tumor microenvironment is a dynamic cellular milieu of a large number of independent components including the immune cells, extracellular matrix, cancer-associated fibroblasts (CAF), cancer stem cells (CSCs), and non-CSCs [33]. Mounting evidence suggests that

TABLE 2 Representative ongoing clinical trials of DNMTis combined with chemotherapy.

NCT number	Recruitment	DNMTis	Chemotherapy	Design	Conditions	Phases
NCT04214860	Recruiting	Azacitidine	APR-246: p53 agonist; Venetoclax: Bcl-2-selective inhibitor	APR-246 + Venetoclax and azacitidine	TP53-mutant myeloid malignancies	Phase 1
NCT03814005	Recruiting	Azacitidine	Pevonedistat: Nedd8-activating enzyme inhibitor; Docetaxel or Paclitaxel: antimicrotubule agent; Carboplatin: alkylating agents	Pevonedistat + azacitidine, docetaxel or paclitaxel + carboplatin	Higher-risk myelodysplastic syndromes, CML, AML, or advanced solid tumors	Phase 1
NCT04172844	Recruiting	Azacitidine	Venetoclax; Pevonedistat	Pevonedistat + azacitidine + venetoclax	AML	Phase 1
NCT03572387	Recruiting	Azacitidine	ATRA: all-trans retinoic acid	Azacitidine + ATRA	Prostate cancer	Phase 2
NCT03542266	Active, not recruiting	Azacitidine	CHOP: cyclophosphamide + doxorubicin hydrochloride + vincristine + prednisone	Azacitidine +CHOP	Untreated peripheral T-cell lymphoma	Phase 2
NCT02250326	Active, not recruiting	Azacitidine	Durvalumab: anti-PD-L1 monoclonal antibody; paclitaxel	Paclitaxel + Azacitidine or durvalumab	Advanced nonsmall cell lung cancer	Phase 2
NCT03683433	Recruiting	Azacitidine	Enasidenib: IDH2 Inhibitor	Azacitidine + Enasidenib	AML with IDH2 gene mutation	Phase 3
NCT03173248	Recruiting	Azacitidine	Ivosidenib: IDH1 inhibitor	Lvosidenib + Azacitidine	Untreated AML with IDH1 mutation	Phase 3
NCT03875287	Recruiting	Decitabine	Cedazuridine: Cytidine Deaminase Inhibitor	Cedazuridine + decitabine	Solid tumor	Phase 1
NCT01567059	Completed	Decitabine	Tosedostat: inhibitor of the M1 family of amino peptidases; Cytarabine: a nucleoside analog	Tosedostat + Cytarabine or Decitabine	AML or high-risk myelodysplastic syndrome	Phase 2
NCT04113616	Recruiting	Decitabine	KRT-232: inhibitor of MDM2; cytarabine	KRT-232 + low-dose cytarabine or decitabine	AML|AML secondary to myeloproliferative neoplasms	Phase 1/2
NCT03709550	Not yet recruiting	Decitabine	Enzalutamide: androgen receptor inhibitors	Decitabine + enzalutamide	Metastatic castration resistant prostate cancer	Phase 1/2

NCT03553537	Not yet recruiting	Decitabine	CHOP	Decitabine + CHOP vs CHOP	Untreated peripheral T-cell lymphoma	Phase 3
NCT02085408	Active, not recruiting	Decitabine	Clofarabine: a purine nucleoside analog; daunorubicin hydrochloride: the hydrochloride salt of an anthracycline antineoplastic antibiotic	Clofarabine or daunorubicin hydrochloride and cytarabine followed by decitabine or observation	Newly diagnosed AML	Phase 3
NCT01627041	Active, not recruiting	Decitabine	Cytarabine; daunorubicin hydrochloride	Decitabine + daunorubicin hydrochloride + cytarabine	AML	Phase 2
NCT01896856	Completed	Guadecitabine	Regorafenib: kinase inhibitors; TAS-102: combination agent of trifluridine and tipiracil hydrochloride	Guadecitabine + irinotecan versus regorafenib or TAS-102	Metastatic colorectal cancer	Phase 1/2
NCT01696032	Completed	Guadecitabine	Carboplatin	Guadecitabine + carboplatin	Ovarian cancer	Phase 2
NCT03913455	Recruiting	Guadecitabine	Carboplatin	Guadecitabine + carboplatin	Extensive-stage small cell lung cancer	Phase 2
NCT00996060	Completed	Hydralazine	Valproic acid: anticonvulsant	Hydralazine + valproic acid	Advanced solid tumor malignancies	Phase 1
NCT00404508	Completed	Hydralazine	Magnesium valproate: anticonvulsant	Hydralazine + magnesium valproate	Chemotherapy resistant solid tumors	Phase 2
NCT00404326	Completed	Hydralazine	Magnesium valproate; Cisplatin: alkylating agents	Hydralazine and magnesium valproate + cisplatin chemoradiation	Cervical cancer	Phase 2
NCT02446652	N/A	Hydralazine	TRANSKRIP: hydralazine/ magnesium; carboplatin	TRANSKRIP + carboplatin and paclitaxel	Recurrent-persistent cervical cancer	Phase 3
NCT00533299	N/A	Hydralazine	Valproate: the magnesium salt of valproic acid	Hydralazine + valproate	Ovarian cancer	Phase 3

abnormal epigenetic changes, especially aberrant DNA methylation in components of the tumor microenvironment, contribute to therapeutic resistance, and treatment with demethylating genes could reverse the methylation pattern in tumor growth-related and/or tumor microenvironment-associated genes. CD157 and CD252 are important regulators of effector cytotoxic T-cells activity, which were shown to be suppressed by DNA methylation in drug-resistant ovarian cancer cells. This immunosuppressive mechanism helps cancer cells to escape the immune response stimulated by chemotherapy, resulting in a decrease in the sensitivity of chemotherapy drugs [34]. In addition, Fang and his colleges analyzed recurrent platinum-resistant ovarian cancer specimens and found that DNA methyltransferase inhibitor guadecitabine could sensitize ovarian cancer cells to platinum drugs, in large part by altering multiple gene expression related to DNA repair and even immune activation [35]. Furthermore, the low-dose histone deacetylase inhibitor entinostat when combined with the DNA methyltransferase inhibitor azacitidine was reported to disrupt the premetastatic niche by inhibiting the trafficking of myeloid-derived suppressor cells (MDSCs) through the downregulation of CCR2 and CXCR2, and promote MDSC differentiation into a more interstitial macrophage-like phenotype, eventually increasing overall survival as compared with conventional chemotherapy [36].

CSCs are recognized as a small proportion of relatively quiescent cells in the tumor microenvironment with self-renewal and unlimited proliferation capability. Growing evidence suggests that CSCs are not only responsible for tumor initiation and progression, but also act as important factors to induce drug resistance. Interestingly, the expression of multiple crucial stem cell factors is regulated by methylation regulation, which indicates that the methylation level could influence drug sensitivity. Indeed, various studies demonstrate that DNMTis have the potential to induce the differentiation or apoptosis of tumor precursor cells to decrease the cancer stem cell population and promote the antitumor effects of chemotherapy [37]. For example, Wu et al. found that low-dose decitabine treatment remarkably enhanced the cytotoxic effects of cisplatin and gemcitabine in basal-like bladder cancer through decrease in genome-wide DNA methylation, gene reexpression, and changes in key cellular regulatory pathways such as STAT3 signaling [38]. Likewise, Wang et al. show that SGI-110 could resensitize the ovarian cancer stemlike cell to platinum chemo drugs through the reexpression of differentiation-associated genes to reduce the stemlike properties of tumor cells [39].

Regional DNA hypermethylation induces chemoresistance

Chemotherapy drugs kill tumor cells by causing DNA damage, cell cycle arrest and promoting cell apoptosis. For example, platinum drugs can cause DNA breaks and inhibit cell mitosis, whereas topoisomerase inhibitors prevent DNA unwinding, cause DNA damage, and induce cell apoptosis. Thus, the expression changes caused by abnormal methylation levels of those gene promoters involved in the abovementioned related pathways may affect the sensitivity of tumor cells to chemotherapy drugs [40–42].

Antiapoptosis proteins and proliferation-related genes

Defects in apoptotic pathways to bypass cell death are common in resistant cancer cells. Among them, aberrant methylation of antiapoptosis genes plays an important role in avoiding apoptosis. For example, Apaf-1 is a cell death effector mediating p53-dependent apoptosis. Soengas et al. found that Apaf-1-negative melanomas were invariably chemoresistant and unable to execute an apoptotic program which could be recovered by treatment with the methylation inhibitor 5-aza-2′-deoxycytidine [43]. Similarly, DNMTis could promote Adriamycin chemosensitivity by upregulating the apoptosis-related protein TRAIL in breast cancer cells [44] and synergize decitabine in tumor cells by inducing caspase-8 and caspase-9 [45].

Insensitivity to antigrowth signals is one of the hallmarks of cancer and also contributes to chemoresistance [46]. Recent studies have shown that aberrant DNA methylation in proliferative genes is a common phenomenon in resistant cancer cells. A high-throughput promoter CpG island microarray analysis conducted in antiestrogen-resistant cells showed that activation of growth-promoting genes, due to promoter hypomethylation, was more frequently observed as compared to gene inactivation by promoter hypermethylation [47]. However, Andrew Stone and his coworkers found that, after long-term treatment with tamoxifen in breast cancer cells, the methylation levels of many antiproliferation-related genes, such as RASAL1, DUSP7, ATP2B4, and GDF15, have also increased, thereby inducing the proliferation of drug-resistant cells and reducing the efficacy of tamoxifen. The addition of 5-Aza was sufficient to demethylate GDF15 expression and sensitize resistant cells to tamoxifen [48].

DNA damage repair proteins

Dysregulated DNA repair pathways decrease the ability of cells to monitor DNA damage and therefore cannot activate the apoptotic response. Additionally, the abnormal increase in gene mutation rate and genome instability can directly or indirectly promote the occurrence of tumor resistance. Evidence has shown that increased DNA methylation in DNA repair genes enhances tumor cell resistance. The DNA repair protein O^6-methylguanine-DNA methyltransferase (MGMT) is known to play a role in the sensitivity to temozolomide. Moen et al. investigated the relationship between MGMT gene body cytosine modification and its gene expression in glioblastoma samples. They found that tumors with the unmethylated MGMT promoter and high gene body methylation maintained a high MGMT expression. Thus, pretreating glioblastoma cell lines with decitabine to reduce MGMT body methylation significantly sensitized them to the temozolomide treatment [49].

Likewise, promoter hypermethylation in another classic DNA repair protein hMLH1 was also demonstrated to render cisplatin resistance in ovarian cancer and 5-fluorouracil in colorectal cancer cell lines, which could be reversed by DNA demethylating agent treatment [50, 51].

Drug-target interactions

It is well known that the prerequisite for the efficacy of chemotherapy is effective drug delivery into the tumor cells to reach an effective inhibitory concentration. Decreased uptake (via downregulation of receptors) and increased drug efflux such as ATP-dependent efflux pumps are a common phenomenon in resistant tumor cells. Accumulating studies support an association of epigenetic changes in drug transporter genes with drug response in cancer, especially DNA methylation [52]. Organic cation transporter (OCT2) was found to be the major transporter that enhances the cellular uptake of oxaliplatin [53]. Liu et al. identified the repression of OCT2 caused by hypermethylated CpG islands and the absence of H3K4 methylation as a potential factor contributing to oxaliplatin resistance in RCC. DNA hypermethylation blocked MYC activation of OCT2 and prevented MYC from recruiting MLL1 to catalyze H3K4me3 at the OCT2 promoter, resulting in repressed OCT2 transcription. Consequently, the addition of decitabine reversed the repression of OCT2 and sensitized RCC cells to oxaliplatin [54]. Metabolic enzymes are also regarded as an important factor to influence the chemotherapy clinical outcome. For example, the UGT1A1 gene encodes an enzyme that acts as a drug detoxifier to reduce irinotecan toxicity. However, UGT1A1 expression is repressed by DNA methylation in some colon cancers. Thus, treatment with demethylating agents could reverse aberrant hypermethylation of UGT1A1 and promote irinotecan clinical outcome [55].

PI3K/Akt pathway

The hyperactivation of the Phosphatidylinositol-3-kinase (PI3K)/Akt/mammalian target of rapamycin (mTOR) signaling is implicated in chemoresistance and cancer progression. Evidence indicates that demethylating agents could act as a chemosensitizer by restoring abnormal hypermethylated genes involved in the Akt-mTOR signaling pathway in various cancers. For example, hypermethylation in the promoter region of the tumor suppressor gene HIN-1 is found in paclitaxel-resistant OCCC. 5-aza-2-dC can restore the expression of HIN-1 in resistant cells and inhibit tumor growth through the HIN-1-AKT-mTOR signaling pathway [56]. In addition, PTEN, a well-known tumor suppressor gene involved in PI3K/Akt/mTOR signaling, was also reported to be inactivated by promoter methylation other than b mutation and homozygous deletion in several types of human tumors [57]. Hypermethylated PTEN was shown to be associated with chemoresistance [58, 59]. This observation indicates the application prospect of demethylating drugs to reverse hypermethylation of PTEN in chemotherapy.

MicroRNAS

MicroRNAs are small noncoding RNAs 19–25 nucleotides in size involved in many biological processes such as survival, apoptosis, cell cycle, and gene regulation [60, 61]. There is an emerging role of microRNAs in resistance to cancer treatments, especially after epigenetic modifications [62]. For instance, the MiR-654 promoter region was hypermethylated in ovarian cancer cell lines and tissues, which was associated with poorer prognosis, and overexpression of miR-654-5p levels reduces cell viability and sensitizes cells to paclitaxel [63]. Results from other studies performed on platinum-resistant ovarian tumors showed that miR-7 methylation is a potential predictive biomarker for recurrence in ovarian cancer patients treated with platinum, and the silencing of miR-7 expression increased levels of its target MAFG and strongly promoted the resistance to platinum in sensitive cells, which can be reverted by combination of demethylating drugs and histone deacetylase (HDAC) inhibitors [64].

Combination of DNMTis with other therapies

DNA methyltransferase inhibitors are effective in sensitizing chemoresistance and its combination with other chromatin or immunological drugs still has broad application prospects in cancer treatment. Immune checkpoint blockade therapy has shown great promise.

However, the majority of patients with cancer develop inevitable resistance. DNMTi has been validated to be an effective sensitizer to immune checkpoint inhibitor (ICI) treatment in multiple cancers. For instance, a recent report indicated that DNMTi treatment could sensitize anti-CTLA4 therapy through upregulating immune signaling by removing methylation from promoter regions of silenced endogenous retroviruses and activating the viral defense pathway in a preclinical melanoma model [65]. In addition, many tumor antigen genes such as the Cancer Testis Antigens (CTAs) are found to remain DNA methylated and silenced in various cancer cells, limiting the immunotherapeutic efficiency, which could be facilitated by reversing the hypermethylation state [66].

Besides, epigenetic regulation during tumorigenesis is complicated and involves multiple steps. Therefore, the combination of two or more therapies targeting various epigenetic events could exert a more prominent antitumor effect. For instance, Huang et al. found that cotreatment of histone lysine demethylase inhibitor oligoamines and DAC suppressed the WNT signaling pathway through inducing the reexpression of the aberrantly silenced WNT signaling pathway antagonist gene, secreted frizzled-related proteins (SFRPs) [67]. Notably, the epigenetic treatment affects not only tumor cells themselves but also the tumor microenvironment. Evidence showed that the combination of azacitidine and ITF-2357 (HDACi) induces profound drug synergy in NSCLC. Azacitidine reactivates some tumor suppressor genes through demethylation while HDACi suppresses histone deacetylases that are involved in processes such as cell proliferation and promotes cancer production, consequently triggering a chemical cascade that recruits more immune cells and reduces the effect of the oncogene MYC.

Conclusions and prospects

DNMTis have been widely applied in various hematological malignancies and some solid tumors. Combining DNMTis with other therapies holds promising prospects. However, attention still deserves to be paid to the disadvantages and limitations, which are listed as follows: Firstly, from the existing studies, there is no reported significant relationship between the DNA methylation/demethylation state and clinical effect. Meanwhile, although multiple preclinical research groups identified biomarkers for DNMTi application, there are no reliable biomarkers to predict DNMTi effects in clinic. Secondly, the role of altered DNA methylation in cancer remains unclearly understood. Aberrant DNA methylation indeed influences chemoresistance, but whether hypermethylation or demethylation causes resistance remains not clearly clarified. Hypermethylation in some models promotes cancer resistance, while in others the situation is reversed. For example, Reduced Folate carrier (RFC) has been recognized to promote methotrexate Purine discharge. Worm et al. found that methotrexate-resistant MCF-7 breast cancer cells harbored inherent defects in drug intake and lack of RFC expression due to hypermethylation in the promoter. DNA methylation inhibitor (DAC) treatment not only increased methotrexate absorption, but also restored reduced folate carrier (RFC) expression. Besides, expression levels of several ATP-dependent efflux pumps, including P-glycoprotein and members of the multidrug resistance-associated protein family, were also increased in response to DAC treatment. These results indicated that while DAC counteracts some methotrexate resistance mechanisms, it also improves other resistance mechanisms [68].

Furthermore, DNMTis demethylation function is not specific and widespread, which means it could reactive oncogenes or silence tumor suppressor genes. Thirdly, the inherent cytotoxic side effects and chemical instability of DNA nucleoside analogues limit their clinical applications for solid tumor featured with a relatively lower proliferative index. Finally, there are still many misunderstandings in the rational use of drugs including choosing the appropriate dose and course of treatment. The DNA methylation process is reversible and demethylating drugs cannot remove cancer stem cells; they must be administered continuously and once the drug is stopped, the disease may relapse. Thus, further development of exploring new types of DNMTis and further research focusing on identifying the detailed mechanisms are needed.

Acknowledgments

We thank the editors and reviewers for their kind support. We especially thank Chunfu Wu and Lihui Wang for their direction and the opportunity to write this volume.

References

[1] Jones PA, Baylin SB. The epigenomics of cancer. Cell 2007;128(4):683–92.

[2] Mohammad HP, Barbash O, Creasy CL. Targeting epigenetic modifications in cancer therapy: erasing the roadmap to cancer. Nat Med 2019;25(3):403–18. https://doi.org/10.1038/s41591-019-0376-8.

[3] Jurkowska RZ, Jurkowski TP, Jeltsch A. Structure and function of mammalian DNA methyltransferases. ChemBioChem 2011;12(2):206–22. https://doi.org/10.1002/cbic.201000195.

[4] Andreu-Vieyra CV, Liang G. Nucleosome occupancy and gene regulation during tumorigenesis. Adv Exp Med Biol 2013;754:109–34. https://doi.org/10.1007/978-1-4419-9967-2_5.

[5] Topper MJ, Vaz M, Marrone KA, Brahmer JR, Baylin SB. The emerging role of epigenetic therapeutics in immuno-oncology. Nat Rev Clin Oncol 2020;17(2):75–90. https://doi.org/10.1038/s41571-019-0266-5.

[6] Yu J, Xie T, Wang Z, Wang X, Zeng S, Kang Y, et al. DNA methyltransferases: emerging targets for the discovery of inhibitors as potent anticancer drugs. Drug Discov Today 2019;24(12):2323–31. https://doi.org/10.1016/j.drudis.2019.08.006.

[7] Gonzalez-Fierro A, Dueñas-González A. Emerging DNA methylation inhibitors for cancer therapy: challenges and prospects. Exp Rev Precis Med Drug Develop 2019;4(1):27–35. https://doi.org/10.1080/23808993.2019.1571906.

[8] Kaiser S, Jurkowski TP, Kellner S, Schneider D, Jeltsch A, Helm M. The RNA methyltransferase Dnmt2 methylates DNA in the structural context of a tRNA. RNA Biol 2017;14(9):1241–51. https://doi.org/10.1080/15476286.2016.1236170.

[9] Lyko F. The DNA methyltransferase family: a versatile toolkit for epigenetic regulation. Nat Rev Genet 2018;19(2):81–92. https://doi.org/10.1038/nrg.2017.80.

[10] Cheng Y, He C, Wang M, Ma X, Mo F, Yang S, et al. Targeting epigenetic regulators for cancer therapy: mechanisms and advances in clinical trials. Signal Transduct Target Ther 2019;4(1):62. https://doi.org/10.1038/s41392-019-0095-0.

[11] Ley TJ, Ding L, Walter MJ, McLellan MD, Lamprecht T, Larson DE, et al. DNMT3A mutations in acute myeloid leukemia. N Engl J Med 2010;363(25):2424–33. https://doi.org/10.1056/NEJMoa1005143.

[12] Abdel-Wahab O, Levine RL. Mutations in epigenetic modifiers in the pathogenesis and therapy of acute myeloid leukemia. Blood 2013;121(18):3563–72. https://doi.org/10.1182/blood-2013-01-451781.

[13] Yan X-J, Xu J, Gu Z-H, Pan C-M, Lu G, Shen Y, et al. Exome sequencing identifies somatic mutations of DNA methyltransferase gene DNMT3A in acute monocytic leukemia. Nat Genet 2011;43(4):309–15. https://doi.org/10.1038/ng.788.

[14] Odejide O, Weigert O, Lane AA, Toscano D, Lunning MA, Kopp N, et al. A targeted mutational landscape of angioimmunoblastic T-cell lymphoma. Blood 2014;123(9):1293–6. https://doi.org/10.1182/blood-2013-10-531509.

[15] Challen GA, Sun D, Jeong M, Luo M, Jelinek J, Berg JS, et al. Dnmt3a is essential for hematopoietic stem cell differentiation. Nat Genet 2012;44(1):23–31. https://doi.org/10.1038/ng.1009.

[16] Guryanova OA, Lieu YK, Garrett-Bakelman FE, Spitzer B, Glass JL, Shank K, et al. Dnmt3a regulates myeloproliferation and liver-specific expansion of hematopoietic stem and progenitor cells. Leukemia 2016;30(5): 1133–42. https://doi.org/10.1038/leu.2015.358.

[17] Ding L, Ley TJ, Larson DE, Miller CA, Koboldt DC, Welch JS, et al. Clonal evolution in relapsed acute myeloid leukaemia revealed by whole-genome sequencing. Nature 2012;481(7382):506–10. https://doi.org/10.1038/nature10738.

[18] Welch JS, Ley TJ, Link DC, Miller CA, Larson DE, Koboldt DC, et al. The origin and evolution of mutations in acute myeloid leukemia. Cell 2012;150(2):264–78. https://doi.org/10.1016/j.cell.2012.06.023.

[19] Issa JP. DNA methylation as a therapeutic target in cancer. Clin Cancer Res 2007;13(6):1634–7. https://doi.org/10.1158/1078-0432.CCR-06-2076.

[20] Jones PA, Issa J-PJ, Baylin S. Targeting the cancer epigenome for therapy. Nat Rev Genet 2016;17(10): 630–41. https://doi.org/10.1038/nrg.2016.93.

[21] Ahuja N, Easwaran H, Baylin SB. Harnessing the potential of epigenetic therapy to target solid tumors. J Clin Invest 2014;124(1):56–63. https://doi.org/10.1172/JCI69736.

[22] Tsai HC, Li H, Van Neste L, Cai Y, Robert C, Rassool FV, et al. Transient low doses of DNA-demethylating agents exert durable antitumor effects on hematological and epithelial tumor cells. Cancer Cell 2012;21(3): 430–46. https://doi.org/10.1016/j.ccr.2011.12.029.

[23] Fang MZ, Wang Y, Ai N, Hou Z, Sun Y, Lu H, et al. Tea polyphenol (−)-epigallocatechin-3-gallate inhibits DNA methyltransferase and reactivates methylation-silenced genes in cancer cell lines. Cancer Res 2003; 63(22):7563–70.

[24] Yoo J, Choi S, Medina-Franco JL. Molecular modeling studies of the novel inhibitors of DNA methyltransferases SGI-1027 and CBC12: implications for the mechanism of inhibition of DNMTs. PLoS One 2013;8(4): e62152. https://doi.org/10.1371/journal.pone.0062152.

[25] Singh N, Dueñas-González A, Lyko F, Medina-Franco JL. Molecular modeling and molecular dynamics studies of hydralazine with human DNA methyltransferase 1. ChemMedChem 2009;4(5):792–9. https://doi.org/10.1002/cmdc.200900017.

[26] Stresemann C, Brueckner B, Musch T, Stopper H, Lyko F. Functional diversity of DNA methyltransferase inhibitors in human cancer cell lines. J Cancer Research 2006;66(5):2794–800. https://doi.org/10.1158/0008-5472.CAN-05-2821%.

[27] Chuang JC, Yoo CB, Kwan JM, Li TWH, Liang G, Yang AS, et al. Comparison of biological effects of non-nucleoside DNA methylation inhibitors versus 5-aza-2′-deoxycytidine. J Mol Cancer Therapeut 2005;4(10): 1515–20. https://doi.org/10.1158/1535-7163.MCT-05-0172%.

[28] Rhee I, Bachman KE, Park BH, Jair K-W, Yen R-WC, Schuebel KE, et al. DNMT1 and DNMT3b cooperate to silence genes in human cancer cells. Nature 2002;416(6880):552–6.

[29] Li J, Hao D, Wang L, Wang H, Wang Y, Zhao Z, et al. Epigenetic targeting drugs potentiate chemotherapeutic effects in solid tumor therapy. Sci Rep 2017;7(1):4035. https://doi.org/10.1038/s41598-017-04406-0.

[30] Segura-Pacheco B, Perez-Cardenas E, Taja-Chayeb L, Chavez-Blanco A, Revilla-Vazquez A, Benitez-Bribiesca L, et al. Global DNA hypermethylation-associated cancer chemotherapy resistance and its reversion with the demethylating agent hydralazine. J Transl Med 2006;4(1):32. https://doi.org/10.1186/1479-5876-4-32.

[31] Festuccia C, Gravina GL, D'Alessandro AM, Muzi P, Millimaggi D, Dolo V, et al. Azacitidine improves antitumor effects of docetaxel and cisplatin in aggressive prostate cancer models. Endocr Relat Cancer 2009;16(2):401–13. https://doi.org/10.1677/ERC-08-0130.

[32] Fang F, Munck J, Tang J, Taverna P, Wang Y, Miller DFB, et al. The novel, small-molecule DNA methylation inhibitor SGI-110 as an ovarian cancer chemosensitizer. J Clin Cancer Res 2014;20(24):6504–16. https://doi.org/10.1158/1078-0432.CCR-14-1553%.

[33] Sun Y. Tumor microenvironment and cancer therapy resistance. Cancer Lett 2016;380(1):205–15. https://doi.org/10.1016/j.canlet.2015.07.044.

[34] Cacan E. Epigenetic-mediated immune suppression of positive co-stimulatory molecules in chemoresistant ovarian cancer cells. Cell Biol Int 2017;41(3):328–39. https://doi.org/10.1002/cbin.10729.

[35] Fang F, Cardenas H, Huang H, Jiang G, Perkins SM, Zhang C, et al. Genomic and epigenomic signatures in ovarian cancer associated with resensitization to platinum drugs. J Cancer Res 2018;78(3):631–44. https://doi.org/10.1158/0008-5472.CAN-17-1492%.
[36] Lu Z, Zou J, Li S, Topper MJ, Tao Y, Zhang H, et al. Epigenetic therapy inhibits metastases by disrupting premetastatic niches. Nature 2020;579(7798):284–90. https://doi.org/10.1038/s41586-020-2054-x.
[37] Ahuja N, Sharma AR, Baylin SB. Epigenetic therapeutics: a new weapon in the war against Cancer. Annu Rev Med 2016;67(1):73–89. https://doi.org/10.1146/annurev-med-111314-035900.
[38] Wu M, Sheng L, Cheng M, Zhang H, Jiang Y, Lin S, et al. Low doses of decitabine improve the chemotherapy efficacy against basal-like bladder cancer by targeting cancer stem cells. Oncogene 2019;38(27):5425–39. https://doi.org/10.1038/s41388-019-0799-1.
[39] Wang Y, Cardenas H, Fang F, Condello S, Taverna P, Segar M, et al. Epigenetic targeting of ovarian cancer stem cells. Cancer Res 2014;74(17):4922–36. https://doi.org/10.1158/0008-5472.CAN-14-1022.
[40] Bonito NA, Borley J, Wilhelm-Benartzi CS, Ghaem-Maghami S, Brown R. Epigenetic regulation of the Homeobox gene MSX1 associates with platinum-resistant disease in high-grade serous epithelial ovarian cancer. Clin Cancer Res 2016;22(12):3097–104. https://doi.org/10.1158/1078-0432.CCR-15-1669.
[41] de Leon M, Cardenas H, Vieth E, Emerson R, Segar M, Liu Y, et al. Transmembrane protein 88 (TMEM88) promoter hypomethylation is associated with platinum resistance in ovarian cancer. Gynecol Oncol 2016;142(3):539–47. https://doi.org/10.1016/j.ygyno.2016.06.017.
[42] Staub J, Chien J, Pan Y, Qian X, Narita K, Aletti G, et al. Epigenetic silencing of HSulf-1 in ovarian cancer:implications in chemoresistance. Oncogene 2007;26(34):4969–78.
[43] Soengas MS, Capodieci P, Polsky D, Mora J, Esteller M, Opitz-Araya X, et al. Inactivation of the apoptosis effector Apaf-1 in malignant melanoma. Nature 2001;409(6817):207–11. https://doi.org/10.1038/35051606.
[44] Xu J, Zhou J-Y, Tainsky MA, Wu GS. Evidence that tumor necrosis factor-related apoptosis-inducing ligand induction by 5-Aza-2′-deoxycytidine sensitizes human breast cancer cells to adriamycin. Cancer Res 2007;67(3):1203–11.
[45] Fulda S, Debatin KM. 5-Aza-2′-deoxycytidine and IFN-gamma cooperate to sensitize for TRAIL-induced apoptosis by upregulating caspase-8. Oncogene 2006;25(37):5125–33.
[46] Hanahan D, Weinberg RA. Hallmarks of cancer: the next generation. Cell 2011;144(5):646–74. https://doi.org/10.1016/j.cell.2011.02.013.
[47] Fan M, Yan PS, Hartman-Frey C, Chen L, Paik H, Oyer SL, et al. Diverse gene expression and DNA methylation profiles correlate with differential adaptation of breast cancer cells to the antiestrogens tamoxifen and fulvestrant. Cancer Res 2006;66(24):11954–66.
[48] Stone A, Valdés-Mora F, Gee JMW, Farrow L, McClelland RA, Fiegl H, et al. Tamoxifen-induced epigenetic silencing of oestrogen-regulated genes in anti-hormone resistant breast Cancer. PLoS One 2012;7(7). https://doi.org/10.1371/journal.pone.0040466, e40466.
[49] Moen EL, Stark AL, Zhang W, Dolan ME, Godley LA. The role of gene body cytosine modifications in MGMT expression and sensitivity to temozolomide. Mol Cancer Ther 2014;13(5):1334–44. https://doi.org/10.1158/1535-7163.MCT-13-0924.
[50] Plumb JA, Strathdee G, Sludden J, Kaye SB, Brown R. Reversal of drug resistance in human tumor xenografts by 2′-deoxy-5-azacytidine-induced demethylation of the hMLH1 gene promoter. Cancer Res 2000;60(21):6039–44.
[51] Arnold CN, Goel A, Boland CR. Role of hMLH1 promoter hypermethylation in drug resistance to 5-fluorouracil in colorectal cancer cell lines. Int J Cancer 2003;106(1):66–73.
[52] Ivanov M, Kacevska M, Ingelman-Sundberg M. Epigenomics and interindividual differences in drug response. Clin Pharmacol Ther 2012;92(6):727–36. https://doi.org/10.1038/clpt.2012.152.
[53] Yokoo S, Yonezawa A, Masuda S, Fukatsu A, Katsura T, Inui K. Differential contribution of organic cation transporters, OCT2 and MATE1, in platinum agent-induced nephrotoxicity. Biochem Pharmacol 2007;74(3):477–87. https://doi.org/10.1016/j.bcp.2007.03.004.
[54] Liu Y, Zheng X, Yu Q, Wang H, Tan F, Zhu Q, et al. Epigenetic activation of the drug transporter OCT2 sensitizes renal cell carcinoma to oxaliplatin. Sci Transl Med 2016;8(348):348ra97. https://doi.org/10.1126/scitranslmed.aaf3124.
[55] Gagnon JF, Bernard O, Villeneuve L, Tetu B, Guillemette C. Irinotecan inactivation is modulated by epigenetic silencing of UGT1A1 in colon cancer. Clin Cancer Res 2006;12(6):1850–8. https://doi.org/10.1158/1078-0432.CCR-05-2130.

[56] Ho CM, Huang CJ, Huang SH, Chang SF, Cheng WF. Demethylation of HIN-1 reverses paclitaxel-resistance of ovarian clear cell carcinoma through the AKT-mTOR signaling pathway. BMC Cancer 2015;15:1471–2407. https://doi.org/10.1186/s12885-015-1744-5.

[57] Soria JC, Lee HY, Lee JI, Wang L, Issa JP, Kemp BL, et al. Lack of PTEN expression in non-small cell lung cancer could be related to promoter methylation. Clin Cancer Res 2002;8(5):1178–84.

[58] Yan B, Yin F, Wang QI, Zhang W, Li LI. Integration and bioinformatics analysis of DNA-methylated genes associated with drug resistance in ovarian cancer. Oncol Lett 2016;12(1):157–66. https://doi.org/10.3892/ol.2016.4608.

[59] Dai FR, Zhang Y, Zhu X, Shan NC, Chen YX. Anticancer role of MUC1 aptamer-miR-29b chimera in epithelial ovarian carcinoma cells through regulation of PTEN methylation. Target Oncol 2012;7(4):217–25. https://doi.org/10.1007/s11523-012-0236-7.

[60] Bartel DP. MicroRNAs: genomics, biogenesis, mechanism, and function. Cell 2004;116(2):281–97. https://doi.org/10.1016/s0092-8674(04)00045-5.

[61] Calin GA, Croce CM. MicroRNA signatures in human cancers. Nat Rev Cancer 2006;6(11):857–66. https://doi.org/10.1038/nrc1997.

[62] He DX, Gu XT, Jiang L, Jin J, Ma X. A methylation-based regulatory network for microRNA 320a in chemoresistant breast cancer. Mol Pharmacol 2014;86(5):536–47. https://doi.org/10.1124/mol.114.092759.

[63] Majem B, Parrilla A, Jimenez C, Suarez-Cabrera L, Barber M, Marin A, et al. MicroRNA-654-5p suppresses ovarian cancer development impacting on MYC, WNT and AKT pathways. Oncogene 2019;38(32):6035–50. https://doi.org/10.1038/s41388-019-0860-0.

[64] Vera O, Jimenez J, Pernia O, Rodriguez-Antolin C, Rodriguez C, Sanchez Cabo F, et al. DNA methylation of miR-7 is a mechanism involved in platinum response through MAFG overexpression in cancer cells. Theranostics 2017;1838–7640. https://doi.org/10.7150/thno.20112.

[65] Chiappinelli Katherine B, Strissel Pamela L, Desrichard A, Li H, Henke C, Akman B, et al. Inhibiting DNA methylation causes an interferon response in cancer via dsRNA including endogenous retroviruses. Cell 2015; 162(5):974–86. https://doi.org/10.1016/j.cell.2015.07.011.

[66] Chiappinelli KB, Zahnow CA, Ahuja N, Baylin SB. Combining epigenetic and immunotherapy to combat cancer. Cancer Res 2016;76(7):1683–9. https://doi.org/10.1158/0008-5472.CAN-15-2125.

[67] Huang Y, Stewart TM, Wu Y, Baylin SB, Marton LJ, Perkins B, et al. Novel oligoamine analogues inhibit lysine-specific demethylase 1 and induce reexpression of epigenetically silenced genes. Clin Cancer Res 2009;15(23): 7217–28. https://doi.org/10.1158/1078-0432.CCR-09-1293.

[68] Worm J, Kirkin AF, Dzhandzhugazyan KN, Guldberg P. Methylation-dependent silencing of the reduced folate carrier gene in inherently methotrexate-resistant human breast cancer cells. J Biol Chem 2001;276(43): 39990–40000. https://doi.org/10.1074/jbc.M103181200.

CHAPTER

3

Histone deacetylase inhibitors (HDACi), the ongoing epigenetic agents to enhance chemotherapy sensitization

Huanran Sun, Jiaqi Song, Mingming Sun, and Changliang Shan

College of Pharmacy, Nankai University, Tianjin, People's Republic of China

Abstract

Epigenetic alterations play a pivotal role in tumor initiation and progression. Histone deacetylases (HDACs) are a family of epigenetic enzymes that affect gene transcription by inducing histone deacetylation, and then regulate fundamental cellular processes including proliferation, differentiation, and development. It is reported that the expression of histone deacetylase (HDAC) is abnormal in multiple cancer. Targeting HDAC to develop inhibitors is an effective way in treating cancer and enhancing sensitization of tumor to chemotherapeutic drugs. In this chapter, we mainly summarize the different classes of histone deacetylase inhibitors (HDACi) and discuss how HDACi combine with chemotherapeutic drugs to enhance their sensitization.

Abbreviations

5-FU	5-fluorouracil
ABHA	azelaicbis-hydroxamic acid
AP	apicidin
AR-FL	full-length AR
AR-Vs	AR splice variants
CBHA	m-carboxycinnamic acid
CI-994	gemcitabine
CQ	chloroquine
CRPC	castration-resistant prostate cancer
DHA	dihydroartemisinin
DOCA	deoxycorticosterone acetate
ERCC1	excision repair cross-complementing 1
FK228	romidepsin

Epigenetic Regulation in Overcoming Chemoresistance
https://doi.org/10.1016/B978-0-12-823684-0.00014-1

HATs	histone acetyl transferases
HCC	hepatocellular carcinoma
HDAC	histone deacetylase
HDACi	histone deacetylase inhibitors
ITF2357	givinostat
LAQ824	dacinostat
LBH589	panobinostat
MDS	myelodysplastic syndrome
MM	multiple myeloma
MS-275	entinostat
NAD	nicotinamide adenine dinucleotide
NPM	nucleophosmin
NSCLC	nonsmall cell lung cancer
PSA	prostate-specific antigen
PXD101	belinostat
ROS	reactive oxygen species
SAHA	vorinostat
SB	sodium butyrate
SB939	pracinostat
TPX	trapoxin
TSA	trichostatinA
TubA	tubastatin A
VPA	valproic acid

Conflict of interest

No potential conflicts of interest were disclosed.

Introduction

Histone acetylation and deacetylation determine the transcription of target genes [1]. The process of acetylation and deacetylation are mediated by histone acetyl transferases (HATs) and histone deacetylases (HDACs), respectively [2]. HDACs are a family of epigenetic enzymes that widely exist in eukaryocytes [3]. HDACs induce histone hypoacetylation by removing the acetyl groups on N-terminal lysines of histone proteins, thereby losing the chromatin structure and finally promoting the transcription process [4]. HDACs play an important role in the epigenetic regulation of gene transcription and control of other cellular functions, especially in cancer [5].

Until now, there are 18 known human histone deacetylases, which are divided into four classes based on homology and function. Class I (HDAC1, HDAC2, HDAC3, and HDAC8), Class IIa (HDAC4, HDAC5, HDAC7, and HDAC9), Class IIb (HDAC6 and HDAC10), and Class IV (HDAC11) enzymes are zinc-dependent enzymes, also known as 'classical' HDACs, while Class III HDACs belong to a zinc-independent and nicotinamide adenine dinucleotide (NAD)-dependent sirtuin enzyme family, which are largely unaffected by HDAC inhibitors [6, 7]. Class I HDACs exist in various types of cells. HDAC1 and 2 are only localized in the nucleus due to the lack of nuclear export signals. HDAC3 is located on the plasma membrane and can shuttle between the nucleus and cytoplasm. HDAC8 is significantly present in cytosol. Compared with Class I HDACs, the expression of Class IIa HDACs is controlled in a tissue-specific manner. Most of them are highly expressed in skeletal muscle, heart, brain, and thymus and play a key role in cell differentiation. Class IIa HDACs exist not only in the cytoplasm but also in the nucleus because of their intrinsic import and export signals. Class IIb HDACs are widely expressed and present in the cytosol and nucleus. Mammalian Class III HDACs extensively expressed in the nucleus and cytosole can act as mono-ADP-ribosyl transferases and affect a wide range of cellular processes such as aging, transcription, apoptosis, and inflammation. The only Class IV HDACs is HDAC11, which has been investigated least. HDAC11 is widely expressed in various tissues such as brain, heart, and testis, and is involved in cell development and immune tolerance processes [8].

HDACs are often abnormally expressed in many disorders, including cancer, which in turn affects gene transcription and normal cellular behavior [9]. The unusually rapid progression of many neoplastic carcinomas such as lymphoma, leukemia, prostate cancer, gastric cancer, breast cancer, and colon cancer has been attributed to the abnormal HDAC activity [10]. High expression levels of HDAC1, 2, and 3 have been proved to be associated with poor outcomes in patients with gastric and ovarian cancers, while high expression of HDAC8 is associated with low survival rate and advanced disease in patients with neuroblastoma [9]. Inhibiting the hyperactive HDAC enzymes has become a crucial therapeutic target against the growth of various cancers [10].

Histone deacetylase inhibitors (HDACi) are a group of small molecules that target classical HDACs by inhibiting their activity, and alter the regulation of histone and nonhistone proteins [9, 11]. HDACi increase the acetylation of core histones, leading to an open chromatin confirmation, which is more accessible to DNA-targeting agents [9]. HDACi have many effects on cells, including not only the induction of proliferation arrest, differentiation, and/or apoptosis of tumor cells, but also the interference of tumor angiogenesis and the regulation of immune response. Therefore, HDACi are developing into exciting and potential anticancer drugs for the treatment of solid and hematological malignancies [11–13].

Cancer, a complicated and frightening disease, contributes to a high mortality rate every year all over the world. Although the mortality rate represents a downward trend, a lot of patients still succumb to this disease. Unlimited proliferation and uncontrolled growth are one of the main features of cancer. Surgery, radiation, and drug therapy such as [9] cytotoxic chemotherapy, targeted therapies, and immunotherapy have played a crucial role in curing cancer, improving patient survival, and changing the quality of life. Among all the therapeutic methods, chemotherapy is one of the most universal ways in clinical treatments, but drug resistance as an important element leads to ineffectiveness of chemotherapy. Resistance to chemotherapy can occur prior to drug treatment (primary or innate resistance) or may develop over time following exposure (acquired resistance) [14]. The molecular mechanisms underlying chemotherapy resistance are complex and usually associated with the following features: (1) Autophagy can be a two-edged sword in therapy because it can not only prevent normal tissue from being tumor tissue but can also contribute to drug resistance through different signal pathways. Prevention of autophagy can improve sensitization to chemotherapy [15]. (2) Another mechanism inducing drug resistance is related to reducing drug uptake by tansmembrance proteins, such as increased cisplatin export by 5′-triphosphate (ATP)-binding cassette transporters [16]. (3) Altered repair of drug-induced DNA damage contributes to a part [17]. (4) Epigenetic changes promote cancer resistance as well [18]. (5) Tumor microenvironment and cancer stem cells also play a role in drug resistance [19, 20]. In recent years, scientists demonstrated that combination chemotherapeutics with HDACi could apparently enhance tumor cell sensitization to chemotherapy drugs and present synergistic or additive antitumor effects, which can reduce cell number and impair cell viability to increase chances of survival.

Classification and application of HDACi

Up to now, 20 different HDACi have been evaluated in clinical trials for anticancer therapies. Some HDACi, such as vorinostat (SAHA), romidepsin (FK228), belinostat (PXD101), and panobinostat (LBH589), have been approved by the US FDA for the treatment of various malignancies. The histone deacetylase inhibitor named chidamide has been approved in China [10, 21]. A few structurally diverse HDACi have been identified. HDACi can be classified, according to their chemical structure, into the following categories: hydroxamic acids, cyclic peptides, short-chain fatty acids, and benzamides [11, 22]. The classification and application of HDACi are summarized below.

Hydroxamic acids as HDACi

Hydroxamic acids is the largest class of HDACi, mainly including trichostatinA (TSA), SAHA, LBH589, PXD101, dacinostat (LAQ824), m-carboxycinnamic acid (CBHA), azelaic bis-hydroxamic acid (ABHA), givinostat (ITF2357), and PCI-34051. SAHA, PXD101, and LBH589 are FDA-approved HDACi. The main functional group of these inhibitors is a hydroxamic acid, which can inhibit the catalysis of HDAC enzymes by interacting with the zinc ion. Generally, hydroxamic acid is a pan-HDAC inhibitor of Class I and Class II HDAC enzymes. TSA and SAHA show stronger efficacy against Class I and IIb HDACs than Class IIa HDACs [6].

TSA is a hydroxamic acid derivative isolated from *Streptomyces*, which represents the most studied compounds and reversibly inhibits HDACs at low nanomolar concentration [7, 23]. SAHA is an effective hydroxamic acid-based inhibitor of HDACs approved by the FDA for the treatment of subcutaneous T-cell lymphoma. This agent has been widely studied both in vitro and in ongoing phase I/II studies in patients with solid tumors and hematological malignancies [22, 24]. Studies have shown that SAHA appears to play a synergistic antitumor role in coordination with anticancer drugs, topoisomerase inhibitors, proteasome inhibitors, TRAIL, and imatinib [23]. Apart from treating cancer, SAHA has shown great potential in reducing memory loss and neuroprotective effects in many studies [24]. LBH589 is a very potent histone deacetylase inhibitor of Class I, II, and IV at low nano-molecular concentrations [10]. It was approved for treatment of multiple myeloma in 2015 [6]. Other hydroxamic acid derivatives also exhibit significant antitumor activity. For example, PXD101 was approved by the FDA in 2014 for treating the peripheral T-cell lymphoma [9]. PCI-34051, a specific inhibitor of HDAC8, can induce cell apoptosis in T-cell lymphoma [25]. Studies have shown that it can delay the xenograft growth of cisplatin-resistant ovarian tumor cells and exhibit good antitumor activity [26]. Scriptaid and ITF 2357 currently have a significant influence on clinical applications due to their relatively low toxicity profile and selective antitumor activity [23]. Pracinostat (SB939) is a novel or a hydroxamic acid-based histone deacetylase inhibitor, which acts on Class I, II, and IV HDACs and has better pharmacokinetic, physicochemical, and pharmaceutical properties than other inhibitors such as SAHA. Previous study has revealed that SB939 is partially effective against 4T1Br4 metastasis to brain in breast cancer [27].

Cyclic peptides as HDACi

Cyclic peptides are also efficient HDACi, which contain functional groups and directly interact with the zinc ion at the catalytic site. Cyclic peptide inhibitors mainly include FK228, apicidin (AP), trapoxin (TPX), depudesin, CHAP, spiruchostatin A, and largazole. These inhibitors are characterized by a surface recognition domain containing a macrocycle with hydrophobic amino acids. Cyclic peptides are generally considered to be Class I HDACi, but because of the large structural differences in such inhibitors, the selectivity among them is different. For example, FK228 is more selective for HDAC1 and 2, while AP has a stronger effect on HDAC2 and 3 [6, 23]. FK228, isolated from *Chromobacterium violaceum*, is an anticancer agent with high-efficiency HDAC inhibitory activity. FK288 selectively inhibits HDAC1 and 2, and induces histone acetylation, cell cycle arrest, and differentiation or apoptosis in vitro and in vivo in some human cancer cell lines. At present, it has been approved as a single drug for the treatment of cutaneous T-cell lymphoma and peripheral T-cell lymphoma [2, 10, 28]. TPX is a cyclic tetrapeptide containing an epoxide group capable of irreversibly binding HDACs. As a microbial metabolite, it induces cell cycle arrest, cell differentiation, and reversal of transformed cells' morphology [7, 29]. AP is a fungal metabolite containing a ketone side chain, which can effectively inhibit HDAC2 and 3. AP reportedly suppresses the proliferation of endometrial carcinoma cells [7, 30].

Short-chain fatty acids as HDACi

Compared with the above two types of HDACi, short-chain fatty acids are relatively weak inhibitors but show good penetration into the brain, making them attractive choices for treating brain diseases [6].

Valproic acid (VPA) is derived from valeric acid, obtained from root extracts of valerian flower plant and belongs to HDACi of short-chain fatty acids. VPA has a lot of effects such as anticonvulsant, antiepileptic, and mood stabilization. It is widely used in clinic. Studies have shown that VPA can also modulate the functional activity of innate and adaptive immune cells by regulating key cell signaling molecules [31]. Among natural-derived HDACi, sodium butyrate (SB), a short-chain fatty acid produced by the bacterial fermentation of fibers in the colon is the most investigated compound. SB, as a histone deacetylase inhibitor, can promote programmed cell death in a number of tumor cells in vitro. It has been shown to induce cell cycle arrest and apoptosis in many kinds of tumor cell types [23]. Phenylbutyrate, a fatty acid with HDAC inhibitory activity, has been approved for treatment of urea cycle disorders and is currently being studied in clinical trials of cancer, hemoglobinopathies, motor neuron diseases, and cystic fibrosis. In addition, it is one of the HDACi already tested in clinical trials for the treatment of recurrent malignant gliomas or myelodysplastic syndrome (MDS) [32]. AR-42 is a new type of phenylbutyrate-based histone deacetylase inhibitor with a low nanomolar IC50. AR-42 exerts antitumor effects on various tumors by regulating transcription factors, cancer-associated proteins, and apoptosis-related pathways. It has higher efficiency and antitumor activity than SAHA. Studies have shown that AR-42 can inhibit the growth and metastasis of canine prostate cancer. At present, the clinical evaluation of AR-42 as an anticancer drug is in progress [33].

Benzamides as HDACi

The benzamide class of HDACi includes several compounds, entinostat (MS-275), chidamide, and gemcitabine (CI-994), which seem to have a high potential in the treatment of different types of tumors. MS-275 is an oral, Class I-specific histone deacetylase inhibitor and the first amino-benzamide-based histone deacetylase inhibitor [34, 35]. Saito et al. found that MS-275, a benzamide derivative, shows HDAC inhibitory activity in vitro and in vivo. It exhibits antitumor activity, low toxicity on normal cells, and strong clinical tolerance in hepatocellular carcinoma cells, breast cancer, and colon cancer. In vitro studies have shown that it can also induce differentiation and apoptosis of human leukemia cells [7, 22, 36]. Chidamide is the first approved oral selective histone deacetylase inhibitor in China, and also the first benzamide-based histone deacetylase inhibitor approved for the treatment of relapsed and refractory peripheral T-cell lymphoma [37]. CI-994, another benzamide-based histone deacetylase inhibitor, is relatively selective for Class I HDACs. CI-994's preference for HDAC1 and 3 is at least 100 times that of HDAC6 and 8 [38]. It has been shown to have significant antitumor activity against a wide variety of rat and human tumor models, and is currently undergoing clinical trials in combination with other standard anticancer agents like carboplatin, paclitaxel, capecitabine, or gemcitabine [39]. Because benzamide-based HDACi have selective inhibitory effect on Class I HDACs and good brain penetration, the treatment of central nervous system diseases with benzamide-based HDACi has become a research hotspot [6].

Other HDACi

Tubastatin A (TubA) is a widely used HDAC6 inhibitor, which has significant therapeutic potential in different animal models of neurodegenerative disorders, autoimmune diseases, cardiovascular diseases, and so on [40]. CG200745 is a recently developed histone deacetylase inhibitor, which can inhibit Class I and Class II HDACs. At present, phase II clinical trials are being conducted to study its anticancer effect. Lee et al. demonstrated that CG200745 can also attenuate hypertension by inhibiting cardiac hypertrophy and fibrosis in hypertensive rats induced by deoxycorticosterone acetate (DOCA) [41]. MGCD0103 is a sulfonamide anilide derivative histone deacetylase inhibitor developed by MethylGene (Montreal, Canada). It is a small-molecule–selective histone deacetylase inhibitor with anticancer activity in vitro. The drug has been studied both in solid tumor and hematological malignancies [22] (Table 1).

HDACi enhance chemotherapy sensitization

HDACi enhance sensitization in nonsmall cell lung cancer

According to the statistic given by Rebecca L. Siegel et al., the nonsmall cell lung cancer (NSCLC) accounts for most lung cancer cases, which causes the largest number of cancer-related deaths [42]. Predicted 5-year survival rate of patients with NSCLC is 15.9% and this rate has only marginally improved during the past few decades [43]. In treating NSCLC, cisplatin is frequently used among platinum agents [44]. However, resistance to cisplatin results

TABLE 1 Classification and application of HDACi.

Class	Compound	Clinical status	Disease/physiological function	Reference
Hydroxamic acids	Vorinostat (SAHA)	Approved for subcutaneous T-cell lymphoma; phase I and II	Solid tumors, acute HIV infection, and hematological malignancies	[22, 24]
	Panobinostat (LBH589)	Approved for multiple myeloma; phase I, II, and III	HIV infection, solid tumors, and hematological malignancies	[6]
	Belinostat (PXD101)	Approved for peripheral T-cell lymphoma; phase I and II	Solid tumors and hematological malignancies	[9]
	Trichostatin A (TSA)		Inhibits histone deacetylases	[7]
	Givinostat (ITF2357)	Phase I and II	Duchenne muscular dystrophy, hematological malignancies, and juvenile idiopathic arthritis	
	PCI-34051		Induces cell apoptosis in T-cell lymphoma, delays the xenografts	[25, 26]
	Pracinostat (SB939)	Phase I and II	Solid tumors and hematological malignancies/inhibits 4T1Br4 metastasis to brain in breast cancer	[27]
	Scriptaid		Treats cancer	
Cyclic peptides	Romidepsin (FK228)	Approved for cutaneous T-cell lymphoma; phase I, II, and III	HIV, hematological malignancies (cutaneous T-cell lymphoma and peripheral T-cell lymphoma), and solid tumors/induces histone acetylation, cell cycle arrest, and differentiation or apoptosis	[10, 28]
	Apicidin (AP)		Suppresses the proliferation of endometrial carcinoma cells	[30]
	Trapoxin (TPX)		Induces cell cycle arrest, cell differentiation and reversal of transformed cells morphology	[29]
Short-chain fatty acids	Valproic acid (VPA)	Phase I, II, III, and IV	Bipolar disorder, hematological malignancies and solid tumors/anticonvulsant, antiepileptic and mood stabilization; modulates the functional activity of innate, and adaptive immune cells	[31]
	Sodium butyrate (SB)	Phase II	Shigellosis/ induces cell cycle arrest, apoptosis and promotes programmed cell death in tumor cells	[23]

Continued

TABLE 1 Classification and application of HDACi—cont'd

Class	Compound	Clinical status	Disease/physiological function	Reference
	Phenylbutyrate	Phase I, II, III, and IV	Urea cycle disorder, pulmonary tuberculosis, syrup urine disease, hemoglobinopathies, motor neuron diseases, recurrent malignant gliomas, and myelodysplastic syndrome	[32]
	AR-42	Phase I	Hematological malignancies/regulates transcription factors, cancer-associated proteins, and apoptosis-related pathway	[32]
Benzamides	Entinostat (MS-275)	Phase I and II	Solid tumors and hematological malignancies	
	Chidamide	Approved for peripheral T-cell lymphoma; phase I and II	Hematological malignancies, HIV infection and solid tumors/induces differentiation, and apoptosis of human leukemia cells	[37]
	CI-994	Phase II and III	Solid tumors and hematological malignancies	
Others	Tubastatin A (TubA)		Has therapeutic potential in animal models of neurodegenerative disorders, autoimmune diseases, and cardiovascular diseases	[40]
	CG200745	Phase II	Cancer/attenuates hypertension in hypertensive rat	[41]
	MGCD0103	Phase I and II	Hematological malignancies and solid tumors	[22]

The contents of table refer to https://clinicaltrials.gov/.

in therapeutic failure. Mechanisms behind cancer cells eluding the cytotoxic potential of cisplatin are complicated, which include consistent reductions in the accumulation of cisplatin, impaired apoptotic signal, and declined recognition of inter- and intrastrand DNA adducts [45].

It is reported that the histone deacetylase inhibitor sensitizes ERCC1-high nonsmall cell lung cancer cells to cisplatin [46]. One of the reasons for poor effectiveness of therapeutic drug cisplatin for NSCLCs is the increased expression of excision repair cross-complementing 1 (ERCC1) gene. Mechanistically, histone deacetylase inhibitor acetylates and actives E2F1 to promote the expression of miR-149, which complementarily combines with *ERCC1* and inhibits the expression of ERCC1. Decreased ERCC1 mRNA level apparently recovers cisplatin sensitization in NSCLCs, and this indicates that HDACi combined with cisplatin to aim at ERCC1-high NSCLC cells is a promising and effective method to improve the survival situation.

Moreover, Shrikant Pradhan et al. also illustrated great efficacy in combining HDACi and cisplatin to enhance NSCLC sensitization to chemotherapy [44]. The hypoxic tumor microenvironment could contribute to acquired chemo/radio resistance. In their study, they combined cisplatin with scriptaid (a histone deacetylase inhibitor) to test the effectiveness of treating NSCLCs. And the conclusion indicated that regardless of normoxia or hypoxia, scriptaid has the ability to increase the efficacy of cisplatin with attenuated metastasis and enhanced DNA damage. Cisplatin resistance could be overcome through promoting apoptosis when cancer cells were treated with additional scriptaid. Therefore, the combination of cisplatin and scriptaid can resolve hypoxia-induced resistance and this method has the potential to improve NSCLC.

Apart from overcoming cisplatin resistance, HDACi can also act on reducing resistance to other drugs such as gefitinib [47] and pemetrexed [48] in NSCLC through different mechanisms.

HDACi enhances sensitization in breast cancer

Breast cancer is the most common cancer disturbing women worldwide whose mortality rates have been improved and it is expected to continue to increase in the next 2–7 years. Besides, the median survival in the metastatic patients is extremely low [49]. These rising cancer rates will be disproportionately high in developing countries and are projected to reach 55% increased incidence and 58% greater mortality in less than 20 years [50].

Resistance to apoptosis due to antibreast cancer agents including tubulin-polymerizing agents (e.g., taxotere) leads to the failure of treatment in which Her-2-mediated signaling accounts for a small part [51]. Her-2, an oncogene overexpressed in approximately 25% breast cancers, is an excellent therapeutic target in breast cancer, and inhibited expression of Her-2 will attenuate its antiapoptotic signaling and suppress the Her-2-induced malignant phenotype [52–54]. The intracellular level of Her-2 is depleted by a novel hydroxamic acid analogue histone deacetylase inhibitor LAQ824 mainly through decreasing its mRNA level and promoting its degradation by proteasome in which pAKT and c-Raf-1 are attenuated and the acetylation of heat shock protein Hsp 90 is induced. Combining LAQ824 with taxotere, epothilone B, and gemcitabine could significantly increase apoptosis of breast cancer cells [51].

HDACi also play an important role in increasing cell cycle arrest in different breast cancer cell lines according to the work of Megan A et al. [55]. They demonstrated that combination HDACi (SB, SAHA) with calpeptin can better reduce live cells and arrest MDA-MB-231 and MCF-7, two breast cancer lines in S phase and G1 phase, respectively, thus presenting wide clinical applications.

The novel histone deacetylase inhibitor FA17 could also effectively sensitize breast cancer cells (MCF-7/MTX) and xenograft tumors to methotrexate and reverse drug resistance by suppressing the activity of nucleophosmin (NPM) and drug efflux pumps with PI3K/Akt and a mitochondrial apoptotic pathway [56].

Furthermore, Liang Gao et al. demonstrated that TSA upregulates p21, thus activating FOXO1 to promote apoptosis. Blocking the mammalian target of rapamycin signaling pathway to induce autophagy is another phenomenon when treating cells with TSA. If an autophagy inhibitor such as chloroquine (CQ) is added to cells with TSA, a synergistic anticancer effect will present [57]. Taken together, utilizing HDACi with chemotherapy drugs is a string that may pull patients out from breast cancer, and more possibilities need to be dug on this therapeutic method.

HDACi enhance sensitization in prostate cancer

According to the statistic [42], prostate cancer has the highest estimated number of new cases (20%) and the second highest estimated deaths (10%), which indicates that prostate cancer is still a threat for people although the death rate for prostate cancer dropped by 51% from 1993 to 2016.

Docetaxel has been used as the first-line chemotherapy for castration-resistant prostate cancer (CRPC); however, most patients with CRPC will ultimately develop resistance to this chemotherapy [58]. Treating CRPC with both HDACi and docetaxel showed synergistic antiproliferative effect in CRPC cells. Mechanistically [58], enhanced levels of full-length AR (AR-FL), AR splice variants (AR-Vs), prostate-specific antigen (PSA), and antiapoptotic Bcl-2 proteins and accelerated acetylation and bundling of tubulin, which significantly inhibited the nuclear accumulation of AR in 22Rv1 cells, are detected compared with docetaxel or HDACi alone. Phase I/II trials were also performed to demonstrate restored sensitivity to bicalutamide in a CRPC model when histone deacetylase inhibitor LBH589 was used in combination with continuous bicalutamide [59]. A cyclic tetrapeptide histone deacetylase inhibitor FK228 also displayed greater therapeutic efficacy when combined with gemcitabine [60] or docetaxel [61] in treating hormone refractory prostate cancer and androgen-independent prostate cancer, respectively. Taken together, combination therapy provides an inspiration for the treatment of patients with different prostate cancers.

HDACi enhance sensitization in gastric cancer

Gastric cancer is the third leading cause of cancer deaths and the most lethal cancer in Eastern Asia [62]. The five-year survival rate then ranges between 30% and 35% for men and women. Patient prognosis remains dismal in countries without GC screening programs [63].

Cisplatin, oxaliplatin, and epirubicin are used to bind to DNA to exert their cytotoxic effects on gastric cancer cells, but the clinical application of these drugs has met the problem of drug resistance [64]. Ramchandra Vigay Amnekar et al. demonstrated for the first time that pretreatment of HDACi contributes to enhanced acetylation level, opened chromatin for accumulating more DNA-targeting drugs and cell cycle arrest; therefore, they synergistically prove a therapeutic effect of chemotherapy drugs in a low dose with reduced toxicity [64]. Besides, the combinatorial regimes include valproic acid with cisplatin or oxaliplatin and TSA with epirubicin, which present greatest cytotoxic potential in inducing cell death at low combined doses in gastric cancer cell lines.

HDACi enhance sensitization in liver cancer

Liver cancer is listed as one of the lowest survival cancers and death rates rose from 2012 through 2016 for cancer of the liver [42]. Although surgical resection, liver transplant and chemotherapy are the primary methods of therapy, surgery is only feasible for about 20% of liver cancer cases and chemotherapy for liver cancer is not satisfactory [65]. Novel therapeutic strategies are explored to ameliorate the situation.

Dihydroartemisinin (DHA) can inhibit liver cancer cells and exhibit antitumor activity; further study has shown that DHA in conjunction with HDACi takes effect in reducing mitochondria membrane potential, releasing cytochrome *c* into the cytoplasm, enhancing p53 and Bak, downregulating Mcl-1 and p-ERK, activating caspase 3 and PARP, and inducing apoptosis. Besides, this combination treatment has better efficacy than DHA alone and does execution in vivo as well [65].

5-Fluorouracil (5-FU) and irinotecan are first-line chemotherapy options for other gastrointestinal tumors, but their efficacy on hepatocellular carcinoma (HCC) is not evident. However, histone deacetylase inhibitor SAHA has been proved to sensitize HCCs to 5-FU and irinotecan in vitro. Loss of the mitochondrial transmembrane potential, downregulation of Bcl-2 expression, and activation of caspase-3 but not caspase-8 were observed when combined treatment induced apoptosis [66].

A multifunctional medicine aspirin has been demonstrated that, in conjunction with histone deacetylase inhibitor VPA, synergistically inhibits liver cancer cell proliferation and migration, and promotes apoptosis and cell cycle accompanied with targeting the Hsp70-mediated signaling pathways of COX in liver cancer cells. Moreover, this combination treatment also takes effect in vivo and does not inhibit the growth of the normal liver cells [67]. As aspirin is widely used in clinical, maybe this therapeutic strategy is a promising one in the treatment of cancer.

Furthermore, fenretinide and HDACi TSA or scriptaid can synergistically overexpress RARβ, which directly or indirectly interacts with Nur77, an apoptosis-related protein to increase cell death and reverse fenretinide resistance [68].

HDACi enhance sensitization in multiple myeloma cells

A series of disorders of mature plasma cells leads to multiple myeloma (MM), whose characteristics include clonal proliferation of malignant plasma cells in the bone marrow microenvironment, monoclonal protein in the blood or urine, and associated organ dysfunction [69]. However, relapse and drug resistance occur in virtually all responding patients [70].

However, interactions between proteasome inhibitor bortezomib and one or two clinical HDACi (e.g., NaB or SAHA) contribute to a synergistic effect on increasing mitochondrial injury, caspase activation, and the induction of apoptosis in MM cells [71], suggesting that the strategy combining these agents is worth further investigation in MM.

Combined exposure of multiple myeloma cells to another proteasome inhibitor carfilzomib in conjunction with histone deacetylase inhibitor LBH589 exhibiting a well-tolerated side-effect profile results in a synergistic inhibition of proliferation in MM cells. Experiment results indicated that the coadministration synergistically increased mitochondrial injury, caspase activation, and apoptosis in MM cells and the lethality of this combination treatment displayed a relationship with the reactive oxygen species (ROS) generation and ERK1/2 inactivation [72].

Summary

Cancer development and tumor progression are associated with abnormal epigenetic modification and transcription patterns. HDACs are key regulators of gene expression that act as transcriptional repressors by removing acetyl groups from histones, which are dysregulated in many cancers, making them a therapeutic target for the treatment of cancer [73]. Currently, some HDACi are approved for use in the clinic as anticancer agents; mechanistically, HDACi prevent deacetylation and form an open chromatin confirmation which is easier for DNA-bound agents to access; thus, they are a promising new class of small-molecular therapeutics [9]. Except for these types of cancer mentioned above where HDACi take effect, other cancers, for example, esophageal cancer [74], chronic lymphocytic leukemia [75], mantle cell lymphoma [76], cervical carcinoma [77], and colon cancer [78] can also be inhibited by combination of HDACi and chemotherapy drugs. HDACi generally interact with conventional chemotherapy drugs cisplatin, taxol, 5-FU, gemcitabine, and bortezomib to restore their functions in diverse cancer cells. In fact, chemotherapy drugs and HDACi promote each other to display synergistic efficacy. The functions of HDACi are far more than enhancing the sensitization of chemotherapy. They perform a role in enhancing sensitization of radiation treatment, immunotherapy and targeted therapy as well. According to the research performed by scientists, HDACi enhance sensitization of chemotherapeutic drugs mechanistically through inducing apoptosis or autophagy, promoting differentiation, and arresting the cell cycle. The targets of HDACi are multiple and complex. More combinational options and mechanisms under them need to be unveiled further to bring hope to patients. Studies also indicate that HDACi have therapeutic potential in a wide range of neurological diseases including stroke, Parkinson's Disease, Alzheimer's Disease, and Huntington's Disease as well as psychiatric diseases including depression and schizophrenia. Moreover, finding more HDACi and promoting them into clinical is similarly urgent.

References

[1] Hou Q, Hu K, Liu X, et al. HADC regulates the diabetic vascular endothelial dysfunction by targetting MnSOD. Biosci Rep 2018;38(5), BSR20181042.

[2] Shukla S, Tekwani BL. Histone deacetylases inhibitors in neurodegenerative diseases, neuroprotection and neuronal differentiation. Frontiers in Pharmacology 2020;11:537.

[3] Han G, Liu N, Li C, et al. Discovery of novel fungal Lanosterol 14alpha-demethylase (CYP51)/histone deacetylase dual inhibitors to treat azole-resistant candidiasis. J Med Chem 2020;63(10):5341–59.

[4] Nepali K, Chang TY, Lai MJ, et al. Purine/purine isoster based scaffolds as new derivatives of benzamide class of HDAC inhibitors. Eur J Med Chem 2020;196:112291.

[5] Chen GL, Wang LH, Wang J, et al. Discovery of a small molecular compound simultaneously targeting RXR and HADC: design, synthesis, molecular docking and bioassay. Bioorg Med Chem Lett 2013;23(13):3891–5.

[6] You D, Richardson JR, Aleksunes LM. Epigenetic regulation of multidrug resistance protein 1 and breast cancer resistance protein transporters by histone deacetylase inhibition. Drug Metab Dispos 2020;48(6):459–80.

[7] Thaler F. Current trends in the development of histone deacetylases inhibitors: a review of recent patent applications. Pharmaceutical Patent Analyst 2012;1(1):75–90.

[8] Singh P, Tomar RS, Rath SK. Anticancer potential of the histone deacetylases inhibitor-like effects of flavones, a subclass of polyphenolic compounds: a review. Mol Biol Rep 2015;42(11):1515–31.

[9] Suraweera A, O'byrne KJ, Richard DJ. Combination therapy with histone deacetylases inhibitors (HDACi) for the treatment of Cancer: achieving the full therapeutic potential of HDACi. Front Oncol 2018;8:92.

[10] Srinivas NR. Clinical pharmacokinetics of panobinostat, a novel histone deacetylase (HDAC) inhibitor: review and perspectives. Xenobiotica 2017;47(4):354–68.

[11] Carafa V, Miceli M, Altucci L, et al. Histone deacetylases inhibitors: a patent review (2009-2011). Expert Opin Ther Pat 2013;23(1):1–17.

[12] Hanif M, Arshad J, Astin J, et al. A multitargeted approach in the discovery of an organorhodium anticancer agent based on Vorinostat as a potent histone deacetylases inhibitor. Angew Chem Int Ed 2020;59(34):14609–14.

[13] Lee KH, Choi EY, Kim MK, et al. Inhibition of histone deacetylase activity down-regulates urokinase plasminogen activator and matrix metalloproteinase-9 expression in gastric cancer. Mol Cell Biochem 2010;343(1–2):163–71.
[14] Giaccone G, Pinedo HM. Drug resistance. Oncologist 1996;1(1 & 2):82–7.
[15] Bhat P, Kriel J, Shubha Priya B, et al. Modulating autophagy in cancer therapy: advancements and challenges for cancer cell death sensitization. Biochem Pharmacol 2018;147:170–82.
[16] Zhu H, Luo H, Zhang W, et al. Molecular mechanisms of cisplatin resistance in cervical cancer. Drug Des Devel Ther 2016;10:1885–95.
[17] Rivera E, Gomez H. Chemotherapy resistance in metas;/tatic breast cancer: the evolving role of ixabepilone. Breast Cancer Res 2010;12(Suppl 2):S2.
[18] Lee JW, Yang DH, Park S, Han HK, Park JW, Kim BY, Um SH, Moon EY. Trichostatin A resistance is facilitated by HIF-1α acetylation in HeLa human cervical cancer cells under normoxic conditions. Oncotarget 2017; 9(2):2035–49.
[19] Wang G, Wang JJ, Yin PH, et al. New strategies for targeting glucose metabolism-mediated acidosis for colorectal cancer therapy. J Cell Physiol 2018;234(1):348–68.
[20] Margaryan NV, Hazard-Jenkins H, Salkeni MA, et al. The stem cell phenotype of aggressive breast cancer cells. Cancers 2019;11(3):340.
[21] Pickell Z, Williams AM, Alam HB, et al. Histone deacetylases inhibitors: a novel strategy for neuroprotection and cardioprotection following ischemia/reperfusion injury. J Am Heart Assoc 2020;9(11), e016349.
[22] Garcia-Manero G, Issa JP. Histone deacetylases inhibitors: a review of their clinical status as antineoplastic agents. Cancer Invest 2005;23(7):635–42.
[23] Emanuele S, Lauricella M, Tesoriere G. Histone deacetylases inhibitors: apoptotic effects and clinical implications (review). Int J Oncol 2008;33(4):637–46.
[24] Wong LW, Chong YS, Wong WLE, et al. Inhibition of histone deacetylase reinstates hippocampus-dependent long-term synaptic plasticity and associative memory in sleep-deprived mice. Cereb Cortex 2020;30(7):4169–82.
[25] An P, Chen F, Li Z, et al. HDAC8 promotes the dissemination of breast cancer cells via AKT/GSK-3beta/snail signals. Oncogene 2020;39(26):4956–69.
[26] Acharya MR, Sparreboom A, Venitz J, et al. Rational development of histone deacetylases inhibitors as anticancer agents: a review. Mol Pharmacol 2005;68(4):917–32.
[27] Chen J, Li N, Liu B, et al. Pracinostat (SB939), a histone deacetylases inhibitor, suppresses breast cancer metastasis and growth by inactivating the IL-6/STAT3 signalling pathways. Life Sci 2020;248:117469.
[28] Hoshino I, Matsubara H, Hanari N, et al. Histone deacetylases inhibitor FK228 activates tumor suppressor Prdx1 with apoptosis induction in esophageal cancer cells. Clin Cancer Res 2005;11(21):7945–52.
[29] Furumai R, Komatsu Y, Nishino N, et al. Potent histone deacetylases inhibitors built from trichostatin A and cyclic tetrapeptide antibiotics including trapoxin. Proc Natl Acad Sci U S A 2001;98(1):87–92.
[30] Fakhry H, Miyamoto T, Kashima H, et al. Immunohistochemical detection of histone deacetylases in endometrial carcinoma: involvement of histone deacetylase 2 in the proliferation of endometrial carcinoma cells. Hum Pathol 2010;41(6):848–58.
[31] Rodriguez-Lopez GM, Soria-Castro R, Campillo-Navarro M, et al. The histone deacetylases inhibitor valproic acid attenuates phospholipase Cgamma2 and IgE-mediated mast cell activation. J Leukoc Biol 2020; 108(3):859–66.
[32] Iannitti T, Palmieri B. Clinical and experimental applications of sodium phenylbutyrate. Drugs R&D 2011; 11(3):227–49.
[33] Duan S, Gong X, Liu X, et al. Histone deacetylases inhibitor, AR-42, exerts antitumor effects by inducing apoptosis and cell cycle arrest in Y79 cells. J Cell Physiol 2019;234(12):22411–23.
[34] Hashimoto A, Fukumoto T, Zhang R, et al. Selective targeting of different populations of myeloid-derived suppressor cells by histone deacetylases inhibitors. Cancer Immunology and Immunotherapy 2020;69(9):1929–36.
[35] Marques AEM, Do Nascimento Filho CHV, Marinho Bezerra TM, et al. Entinostat is a novel therapeutic agent to treat oral squamous cell carcinoma. J Oral Pathol Med 2020;49(8):771–9.
[36] Liu T, Guan F, Wang Y, et al. MS-275 combined with cisplatin exerts synergistic antitumor effects in human esophageal squamous cell carcinoma cells. Toxicol Appl Pharmacol 2020;395:114971.
[37] Xu Y, Zhang P, Liu Y. Chidamide tablets: HDAC inhibition to treat lymphoma. Drugs of Today 2017;53(3): 167–76.

[38] Zhang S, Fujita Y, Matsuzaki R, et al. Class I histone deacetylase (HDAC) inhibitor CI-994 promotes functional recovery following spinal cord injury. Cell Death Dis 2018;9(5):460.
[39] Thomas M, Clarhaut J, Tranoy-Opalinski I, et al. Synthesis and biological evaluation of glucuronide prodrugs of the histone deacetylases inhibitor CI-994 for application in selective cancer chemotherapy. Bioorg Med Chem 2008;16(17):8109–16.
[40] Shen S, Svoboda M, Zhang G, et al. Structural and in vivo characterization of Tubastatin a, a widely used histone deacetylase 6 inhibitor. ACS Med Chem Lett 2020;11(5):706–12.
[41] Yoon GE, Jung JK, Lee YH, et al. Histone deacetylases inhibitor CG200745 ameliorates high-fat diet-induced hypertension via inhibition of angiotensin II production. Naunyn Schmiedebergs Arch Pharmacol 2020;393(3):491–500.
[42] Siegel RL, Miller KD, Jemal A. Cancer statistics, 2019. CA Cancer J Clin 2019;69(1):7–34.
[43] Chen Z, Fillmore CM, Hammerman PS, et al. Non-small-cell lung cancers: a heterogeneous set of diseases. Nat Rev Cancer 2014;14(8):535–46.
[44] Pradhan S, Mahajan D, Kaur P, Pandey N, Sharma C, Srivastava T. Scriptaid overcomes hypoxia-induced cisplatin resistance in both wild-type and mutant p53 lung cancer cells. Oncotarget 2016;7(44):71841–55.
[45] Galluzzi L, Senovilla L, Vitale I, et al. Molecular mechanisms of cisplatin resistance. Oncogene 2012;31(15):1869–83.
[46] He Y, Chen D, Yi Y, et al. Histone deacetylases inhibitor sensitizes ERCC1-high non-small-cell lung cancer cells to cisplatin via regulating miR-149. Molecular Therapy Oncolytics 2020;17:448–59.
[47] Chen H, Wang Y, Lin C, et al. Vorinostat and metformin sensitize EGFR-TKI resistant NSCLC cells via BIM-dependent apoptosis induction. Oncotarget 2017;8(55):93825–38.
[48] Martile M, Gabellini C, Monica V, Busso S, Eramo A, De Maria R, Milella M, Trisciuoglio D. Histone deacetylase inhibition synergistically enhances pemetrexed cytotoxicity through induction of apoptosis and autophagy in non-small cell lung cancer. Mol Cancer 2014;13:230.
[49] Anastasiadi Z, Lianos GD, Ignatiadou E, et al. Breast cancer in young women: an overview. Updates Surg 2017;69(3):313–7.
[50] Villarreal-Garza CAC, Magallanes-Hoyos MC, et al. Breast cancer in young women in Latin America: an unmet, growing burden. Oncologist 2013;18(12):1298–306.
[51] Fuino L, Bali P, Wittmann S, et al. Histone deacetylases inhibitor LAQ824 down-regulates her-2 and sensitizes human breast cancer cells to trastuzumab, taxotere, gemcitabine, and epothilone B. Mol Cancer Ther 2003;2(10):971–84.
[52] Slamon Dj GW, Jones LA, et al. Studies of the HER-2/neu proto-oncogene in human breast and ovarian cancer. Science 1989;244(4905):707–12.
[53] Roh HPJ, Green DW, et al. HER2/neu antisense targeting of human breast carcinoma. Oncogene 2000;19(53):6138–43.
[54] Yu DHM. Overexpression of ErbB2 in cancer and ErbB2-targeting strategies. Oncogene 2000;19(53):6115–21.
[55] Mataga MA, Rosenthal S, Heerboth S, et al. Anti-breast cancer effects of histone deacetylases inhibitors and calpain inhibitor. Anticancer Res 2012;32(7):2523–9.
[56] Chen SY, Zheng XW, Cai JX, et al. Histone deacetylases inhibitor reverses multidrug resistance by attenuating the nucleophosmin level through PI3K/Akt pathway in breast cancer. Int J Oncol 2016;49(1):294–304.
[57] Gao L, Sun X, Zhang Q, et al. Histone deacetylases inhibitor trichostatin a and autophagy inhibitor chloroquine synergistically exert anti-tumor activity in H-ras transformed breast epithelial cells. Mol Med Rep 2018;17(3):4345–50.
[58] Park SE, Kim HG, Kim DE, et al. Combination treatment with docetaxel and histone deacetylases inhibitors downregulates androgen receptor signaling in castration-resistant prostate cancer. Invest New Drugs 2018;36(2):195–205.
[59] Ferrari AC, Alumkal JJ, Stein MN, et al. Epigenetic therapy with Panobinostat combined with bicalutamide rechallenge in castration-resistant prostate cancer. Clin Cancer Res 2019;25(1):52–63.
[60] Kanzaki M, Kakinuma H, Kumazawa T, et al. Low concentrations of the histone deacetylases inhibitor, depsipeptide, enhance the effects of gemcitabine and docetaxel in hormone refractory prostate cancer cells. Oncol Rep 2007;17(4):761–7.
[61] Zhang Z, Stanfield J, Frenkel E, et al. Enhanced therapeutic effect on androgen-independent prostate cancer by depsipeptide (FK228), a histone deacetylases inhibitor, in combination with docetaxel. Urology 2007;70(2):396–401.
[62] Bray F, Ferlay J, Soerjomataram I, et al. Global cancer statistics 2018: GLOBOCAN estimates of incidence and mortality worldwide for 36 cancers in 185 countries. CA Cancer J Clin 2018;68(6):394–424.

[63] Rocken C. Molecular classification of gastric cancer. Expert Rev Mol Diagn 2017;17(3):293–301.
[64] Amnekar RV, Khan SA, Rashid M, et al. Histone deacetylases inhibitor pre-treatment enhances the efficacy of DNA-interacting chemotherapeutic drugs in gastric cancer. World J Gastroenterol 2020;26(6):598–613.
[65] Zhang CZ, Pan Y, Cao Y, et al. Histone deacetylases inhibitors facilitate dihydroartemisinin-induced apoptosis in liver cancer in vitro and in vivo. PloS One 2012;7(6), e39870.
[66] Ocker M, Alajati A, Ganslmayer M, et al. The histone-deacetylase inhibitor SAHA potentiates proapoptotic effects of 5-fluorouracil and irinotecan in hepatoma cells. J Cancer Res Clin Oncol 2005;131(6):385–94.
[67] Li X, Zhu Y, He H, et al. Synergistically killing activity of aspirin and histone deacetylases inhibitor valproic acid (VPA) on hepatocellular cancer cells. Biochem Biophys Res Commun 2013;436(2):259–64.
[68] Yang H, Zhan Q, Wan YJ. Enrichment of Nur77 mediated by retinoic acid receptor beta leads to apoptosis of human hepatocellular carcinoma cells induced by fenretinide and histone deacetylases inhibitors. Hepatology 2011;53(3):865–74.
[69] Palumbo A, Anderson K. Multiple myeloma. N Engl J Med 2011;364(11):1046–60.
[70] Chen S, Zhang Y, Zhou L, et al. A Bim-targeting strategy overcomes adaptive bortezomib resistance in myeloma through a novel link between autophagy and apoptosis. Blood 2014;124(17):2687–97.
[71] Pei XY, Dai Y, Grant S. Synergistic induction of oxidative injury and apoptosis in human multiple myeloma cells by the proteasome inhibitor bortezomib and histone deacetylases inhibitors. Clin Cancer Res 2004;10(11): 3839–52.
[72] Gao L, Gao M, Yang G, et al. Synergistic activity of carfilzomib and Panobinostat in multiple myeloma cells via modulation of ROS generation and ERK1/2. Biomed Res Int 2015;2015:459052.
[73] Carew JS, Giles FJ, Nawrocki ST. Histone deacetylases inhibitors: mechanisms of cell death and promise in combination cancer therapy. Cancer Lett 2008;269(1):7–17.
[74] Ahrens TD, Timme S, Hoeppner J, et al. Selective inhibition of esophageal cancer cells by combination of HDAC inhibitors and Azacytidine. Epigenetics 2015;10(5):431–45.
[75] Dai Y, Chen S, Kramer LB, et al. Interactions between bortezomib and romidepsin and belinostat in chronic lymphocytic leukemia cells. Clin Cancer Res 2008;14(2):549–58.
[76] Dasmahapatra G, Lembersky D, Son MP, et al. Carfilzomib interacts synergistically with histone deacetylases inhibitors in mantle cell lymphoma cells in vitro and in vivo. Mol Cancer Ther 2011;10(9):1686–97.
[77] Huang Z, Peng S, Knoff J, et al. Combination of proteasome and HDAC inhibitor enhances HPV16 E7-specific CD8+ T cell immune response and antitumor effects in a preclinical cervical cancer model. J Biomed Sci 2015;22(1):7.
[78] Tumber A, Collins LS, Petersen KD, et al. The histone deacetylases inhibitor PXD101 synergises with 5-fluorouracil to inhibit colon cancer cell growth in vitro and in vivo. Cancer Chemother Pharmacol 2007; 60(2):275–83.

CHAPTER

4

Histone lysine demethylase inhibitor (HDMi) as chemo-sensitizing agent

Ao Shen and Xi-Yong Yu

Key Laboratory of Molecular Target & Clinical Pharmacology and the State Key Laboratory of Respiratory Disease, School of Pharmaceutical Sciences & The Fifth Affiliated Hospital, Guangzhou Medical University, Guangzhou, People's Republic of China

Abstract

The past two decades have brought an increasing awareness that a wide range of cancers show aberrant epigenetic changes which contribute to oncogenesis. Historically, attempts to pharmacologically intervene in epigenetic dysregulation in cancers started off with DNA methylation and histone acetylation, which have been extensively reviewed in Chapters 2 and 3 of the book. More recently, histone demethylases are identified as one of the most frequently dysregulated epigenetic modifiers in a wide range of cancers. This potentially provides a promising therapeutic strategy using histone lysine demethylase inhibitor (HDMi) to treat chemotherapy-resistant cancers. This chapter presents the rationale for targeting histone demethylases in oncology and provides an update on a few key targets and their inhibitors being investigated as chemo-sensitizing agents.

Abbreviations

2-OG	2-oxoglutarate
2,4-PDCA	*para* 2,4 dicarboxylic acid
AML	acute myeloid leukemia
APL	acute promyelocytic leukemia
ATRA	all-trans retinoic acid
CSC	cancer stem cell
FAD	flavin adenine dinucleotide
HCC	hepatocellular carcinoma
HDMi	histone lysine demethylase inhibitor
JmjC	JumonjiC
KDM	histone lysine demethylase
LSD	lysine-specific demethylase
MAO	monoamine oxidase
MAOi	monoamine oxidase inhibitors

Epigenetic Regulation in Overcoming Chemoresistance
https://doi.org/10.1016/B978-0-12-823684-0.00007-4

NOG *N*-oxalylglycine
NSCLC nonsmall cell lung cancer
OS osteosarcoma
SCLC small cell lung cancer
TCP tranylcypromine
TMZ temozolomide

Conflict of interest

No potential conflicts of interest were disclosed.

The nature of histone modifications

Rather than permanently changing the DNA nucleotide sequence, epigenetic alterations work through reversible modifications on DNA, RNA, and histones and also through nucleosome remodeling [1]. Within the cell nucleus, DNA is packaged into chromatin as repeating units referred to as nucleosomes. Each nucleosome comprises 145–147 base pairs of DNA wrapped around an octameric histone core (made up of two copies each of histones H2A, H2B, H3, and H4) [2], along with one copy of histone H1 as the linker. The nucleosomes are then organized in a relatively loose form in euchromatin or densely compacted form in heterochromatin. Importantly, the level of compaction, together with the presence or absence of various posttranslational modifications on histones, has profound effects on genome stability, gene expression, and other DNA metabolic processes [3].

The nucleosome is approximately globular, except for N-terminal tails of H2B, H3, and H4, and the C-terminal tail of H2A protruding from it. On these tails many lysine, arginine, serine, threonine, tyrosine, and histidine residues are subject to covalent modifications, such as phosphorylation, ubiquitination, acetylation, methylation and, to a lesser extent, sumoylation and citrullination [3, 4]. These modifications are essential for maintaining genomic integrity and playing an important role in determining gene expression. Additional types of histone modifications have been documented including, but not limited to, formylation, propionylation, serotonylation, malonylation, and succinylation, though much is still unknown about the impact each modification has on gene expression and chromatin structure and function.

The dynamics of histone methylation

Histone methylation mainly occurs on the side chains of certain solvent-accessible lysines (K) and arginines (R) on the H3 and H4 tails. Unlike acetylation and phosphorylation altering amino acid charges, methylation increases histone hydrophobicity. Hydrophobic methyl groups (me) can be sequentially added and result in monomethylated, dimethylated, or trimethylated lysines, or result in monomethylated, symmetrically dimethylated or asymmetrically dimethylated arginines. A conventional code has been developed to indicate the site and status of histone methylation. The substrate is specified by the histone subunit and an IUPAC code for amino acid and the number of the amino acid that is modified. The states of methylation are also noted by the addition of "me#," with the numbers being 1, 2, and 3 for monomethylated, dimethylated, and trimethylated substrates, respectively. For example, H3K9me1 means histone H3 with a monomethylated lysine, which is the ninth amino acid of H3.

The most studied lysine methylation and demethylation sites are K4, K9, K27, K36, K79 on histone H3 and K20 on histone H4 (Table 1). Other methylated lysines are also found

TABLE 1 Systematic nomenclature of histone lysine demethylases and their histone substrates.

Subfamily	Symbols	Aliases	Primary target site	Substrates	Secondary target site	Substrates	Additional site(s)
KDM1	KDM1A	LSD1, AOF2, BHC110	H3K4	Me1, Me2	H3K9	Me1, Me2	
	KDM1B	LDS2, AOF1	H3K4	Me1, Me2			
KDM2	KDM2A	JHDM1A, FBXL11	H3K36	Me1, Me2	H3K4	Me3	
	KDM2B	JHDM1B, FBXL10	H3K36	Me1, Me2	H3K79	Me2, Me3	
KDM3	KDM3A	JMJD1A, JHDM2A, TSGA	H3K9	Me1, Me2			
	KDM3B	JMJD1B, JHDM2B	H3K9	Me1, Me2			
	JMJD1C	KDM3C, JHDM2C, TRIP8	H3K9	Me1, Me2			
KDM4	KDM4A	JMJD2A, JHDM3A	H3K9	Me2, Me3	H3K36	Me3	H1.4K26, H3K27
	KDM4B	JMJD2B, JMDM3B	H3K9	Me2, Me3	H3K36	Me3	H1.4K26, H3K27
	KDM4C	JMJD2C, JMDM3C, GASC1	H3K9	Me2, Me3	H3K36	Me3	H1.4K26
	KDM4D	JMJD2D, JMDM3D	H3K9	Me2, Me3			H1.4K26, H3K27
	KDM4E	JMJD2E	H3K9	Me2, Me3			
KDM5	KDM5A	JARID1A, RBP2	H3K4	Me2, Me3			
	KDM5B	JARID1B, PLU-1, FBXL	H3K4	Me2, Me3			
	KDM5C	JARID1C, SMCX	H3K4	Me2, Me3			
	KDM5D	JARID1D, SMCY	H3K4	Me2, Me3			

Continued

TABLE 1 Systematic nomenclature of histone lysine demethylases and their histone substrates—cont'd

Subfamily	Symbols	Aliases	Primary target site	Substrates	Secondary target site	Substrates	Additional site(s)
KDM6	KDM6A	UTX	H3K27	Me2, Me3			
	KDM6B	JMJD3	H3K27	Me2, Me3			
	UTY	KDM6C	H3K27	Me2, Me3			
KDM7	KDM7A	JHDM1D	H3K9	Me1, Me2	H3K27	Me1, Me2	H4K20Me1
	PHF8	KDM7B, JHDM1F	H3K9	Me1, Me2	H3K20	Me1	
	PHF2	KDM7C, JHDM1E	H3K9	Me1, Me2	H3K27	Me1, Me2	H4K20Me3
KDM8	KDM8	JMJD5	H3K36	Me2			
Other	RIOX1	JMJD9, NO66	H3K4	Me2, Me3	H3K36	Me2, Me3	
	RIOX2	JMJD10, MINA	H3K27	Me2, Me3	H3K9me3		

in H1, H2A, H2B and additional sites in H3 and H4. While methylations at some sites are mainly associated with gene activation (like H3K4 and H3K79) or suppression (like H3K9 and H3K27) [5], more often the outcome of a histone methylation depends on the site of methylation, the state of methylation, as well as other modifications on the same or other histones [6, 7].

Since the epigenetic modifications are not permanent, cellular homeostasis not only needs a precise balance of histone methylation and demethylation, but moreover a fine tune of mono-, di-, and trimethylation. Such meticulous methylation marks are orchestrated by specific enzyme families, which can either transfer the methyl group onto the histone (methyltransferases) commonly termed "writers" or, conversely, remove these methyl groups (demethylases) known as "erasers." Additionally, the proteins that specifically recognize epigenetic patterns and can further recruit different effector proteins are called "readers." Different readers that specifically recognize each individual histone methylation and influence the binding of chromatin-associated proteins have been described elsewhere [8]. Consequently, regulations of these modifications are relevant to both normal development and diseases like cancer.

Histone lysine demethylases

Histone demethylation is largely operated by histone lysine demethylases (KDMs), while the existence of a specific histone arginine demethylase remains to be proven [9, 10]. Since first discovered in 2004, two classes of KDMs have been described: the monoamine oxidase (MAO) homolog lysine-specific demethylases (LSD, also known as KDM1) and the JumonjiC (JmjC) domain-containing demethylases. The latter contains over 30 members and can further be divided into at least seven subfamilies (from KDM2 to KDM8) based on the JmjC-domain homology [11–13] (Table 1). The discoveries of different classes and subfamilies of KDM highlight the specific and dynamic regulation of methylation at different histones and various lysine residues.

Each subfamily of KDM acts on different substrates and plays a different role in cellular function. The target specificities of KDMs are regulated by their participation in different complexes. KDMs also have a variety of domains that serve different functions, which include binding to the histone, recognizing the correct methylated substrate and catalyzing the reaction, and binding cofactors and other interaction partners. Both classes of KDM operate by hydroxylation of a methyl group followed by dissociation of formaldehyde; however, they possess different catalytic mechanisms. The LSD-family KDMs use flavin adenine dinucleotide (FAD) as a cofactor and can only demethylate monomethylated and dimethylated but not trimethylated lysines [11]. The JmjC domain-containing KDMs use Fe(II) and 2-oxoglutarate (2-OG) as cofactors and can demethylate all three methylated lysine states [14–16].

Dysregulation of KDMs in cancers

Chemotherapy resistance is a major challenge for cancer treatment, which leads to high mortality among cancer patients. Chemotherapy often fails to eliminate cancer stem cells (CSCs) that drive cancer recurrence, and more importantly, treated tumors usually contain a higher frequency of chemoresistant CSCs [17, 18]. In the past two decades, emerging evidence suggests that genes regulating epigenetics have been altered in a wide range of cancer types, particularly in CSCs of those cancers [19]. Besides DNA methylation and histone acetylation that have been reviewed in the previous chapters, aberrant histone methylation is thought to influence the pathobiology of cancer by promoting tumor proliferation and invasion, reducing tumor suppressor expression, and acquiring chemoresistance and disease relapse [20–22].

The aberrant histone methylation is mainly due to either regulatory defects or hyperactivities of KDMs and/or histone methyltransferases. KDMs are implicated in many different cancers, including, to name only a few, leukemia, lung cancer, prostate cancer, breast cancer, neuroblastoma, glioblastoma, and esophageal squamous carcinoma [20–22]. Because the overall biological outcomes rely on joint effects of histone methylation with other regulatory processes, KDMs thus can activate or repress gene transcription depending on the cellular context and the surrounding chromatin landscape. In tumors, the dysregulation of KDMs causes different outcomes depending on the tissue of origin and other genetic or epigenetic changes. Although the exact mechanisms are yet to be determined in many cases, the oncogenic effects of KDMs are attributed to direct or indirect regulation of either tumor suppressors or oncogenes, or both.

Recent advances in the cancer genome atlas and cancer proteogenomic projects provide us a comprehensive understanding of both oncogenic and tumor suppressive functions of KDMs [23–28]. For example, in one project, researchers have identified 299 genes that drive tumor progression, including KDM5C and KDM6A [24]. Studies since then have identified KDMs as one of the most frequently dysregulated enzymes in many blood or solid tumors. Almost in each KDM subfamily there are dysregulated KDMs that contribute to oncogenesis. Some recent examples include KDM1A (LSD1), KDM2B, KDM3A, KDM4B, KDM5A/5B, and KDM6A/6B, which will be briefly summarized below.

Aberrant KDMs and chemoresistance

Increased expression levels and activities of KDMs are often in many cancers associated with chemotherapy resistance. For example, KDM1A (LSD1) is overexpressed in many solid tumors and in acute myeloid leukemia (AML) [29, 30]. LSD1 inactivation in vitro leads to increased cell differentiation and reduced cell growth of AML and small cell lung cancer (SCLC) [31, 32]. Acute promyelocytic leukemia (APL), a subtype of AML, is very sensitive to all-trans retinoic acid (ATRA) therapy, making APL the only highly curable AML [33]. However, for non-APL AML patients, ATRA has little effect. In AML, LSD1 is found to repress differentiation markers such as E-cadherin [34]. In a murine leukemia model, increased LSD1 expression promotes MYC expression, leading to CSC proliferation and a loss of normal differentiation [31]. Since LSD1 inhibition results in the differentiation of leukemia cells in vitro, this suggests that LSD1 contributes to ATRA resistance in non-APL AML. Consistently, LSD1 inhibitor sensitizes non-APL AML cells to ATRA [35]. In hepatocellular carcinoma (HCC), inhibition of LSD1 improves chemotherapeutic efficacies against HCC [36]. Mechanistically, LSD1 promotes β-catenin activation by inhibiting the expression of several β-catenin suppressors, especially Prickle1 and APC in $Lgr5^+$ CSCs, by directly regulating the levels of H3K4 at the promoters of these genes, thus promoting the stemness and chemoresistance of $Lgr5^+$ CSCs [36].

Like the LSD family, JmjC-family demethylases are implicated in cancer largely through expression and functional characterization in cancer cells. KDM5A (JARID1A) is the first KDM found to associate with a chemoresistant state in a subpopulation of lung cancer cells [37]. Similarly, inhibition of KDM5B (JARID1B) increases the radiosensitivity of breast cancer cells [38] and cisplatin sensitivity of neuroblastoma cells [39], although the exact KDM5A and 5B targets responsible for those phenotypes are yet to be identified. In glioblastoma, KDM5A is a key factor for the resistance to temozolomide (TMZ), as TMZ resistance or restored TMZ sensitivity can be mimicked by overexpressing or inactivating KDM5A, and this treatment-induced TMZ resistance is due to a subset of slow-cycling and transiently partially differentiated cells that escape drug cytotoxicity, overcome the G2 checkpoint, and sustain clonal growth [40]. In another report, through elevated KDM5B expression, melanoma cells shifted toward a more drug-tolerant $CD34^-$ state upon exposure to the BRAF inhibitor or combined BRAF/MEK inhibitors treatment, while KDM5B inhibition shifted melanoma cells to the more BRAF inhibitor-sensitive $CD34^+$ state [41]. Thus, inhibition of KDM5 can regulate tumor heterogeneity and reduce chemotherapy resistance.

Besides LSD1 and KDM5, other KDM subfamilies also play roles in chemoresistance in many types of cancers. KDM2B is highly expressed in glioblastoma surgical specimens [42]. Inhibiting KDM2B impairs the survival of patient-derived primary glioblastoma cells through the induction of DNA damage and apoptosis, sensitizing them to chemotherapy [42]. KDM3A regulates ovarian cancer stemness and cisplatin resistance by inducing the expressions of pluripotent molecules Sox2 and Nanog and antiapoptotic Bcl-2 [43]. KDM4B promotes prostate cancer cell proliferation by targeting cell cycle regulators such as Plk1, while inhibition of KDM4B alters gene expression and slows down tumor cell growth [44].

Clinical data show that KDM6A/6B are upregulated in osteosarcoma (OS) after cisplatin treatment, and consistently, cisplatin-resistant OS specimens have a lower H3K27me3 level than cisplatin-sensitive specimens. Moreover, inhibition of KDM6A/6B can reverse OS cisplatin resistance both in vitro and in vivo [45]. Mechanistically, KDM6A/6B reduces the H3K27me3 level on stemness-related genes such as PRKCA and MCL1, leading to the activation of these genes, ultimately activating the RAF/ERK/MAPK pathway and antiapoptotic Bcl-2, rendering OS resistant to cisplatin [45]. A similar result is observed in diffuse large B-cell lymphoma, as KDM6B inhibition sensitizes diffuse large B-cell lymphoma to R-CHOP chemotherapy [46]. In a most recent report on colorectal cancer, a low H3K27me3 level predicts a poor outcome upon chemotherapeutic treatment [47]. Mechanistically, oxaliplatin stimulation induces NOTCH2 expression, which promotes the expression of stemness-related genes and results in oxaliplatin resistance. This is done by downregulation of the H3K27me3 level on the NOTCH2 transcription initiation site due to overexpression of KDM6A and 6B [47]. Elevation of the H3K27me3 level through KDM6A and 6B depletion or KDM inhibitor GSK-J4 treatment significantly enhances oxaliplatin-induced apoptosis. Furthermore, the combination of GSK-J4 and oxaliplatin significantly inhibits tumor growth in an oxaliplatin-resistant patient-derived xenograft model [47].

One type of cancer can have dysregulation of multiple KDMs. For example, reports from different groups have shown the upregulation of KDM4A/4B [48], KDM5B [49], and KDM5C [50] in breast cancers. Thus, a pan inhibitor targeting more than one KDM might be used to sensitize resistant tumor cells via synergistically inhibiting multiple dysregulated KDMs at the same time. In ovarian cancer, cisplatin treatment triggers the transdifferentiation of tumor cells into CSCs. A compound GHDM-1515 that acts on KDM pathways significantly enhances the efficacy of cisplatin chemotherapy and induces CSCs apoptosis [51]. In nonsmall cell lung cancer (NSCLC), cells with taxane-platin-resistant, KDM3B, and other KDMs' expression increase and develop hypersensitivity to JIB-04 and GSK-J4, two JmjC inhibitors [52].

Finally, unlike deleterious mutations or translocations of genes encoding methyltransferases (e.g., EZH2, DOT1L), which are often associated with certain types of cancer developments (e.g., lymphomas, leukemia), somatic alterations of genes encoding KDMs in cancers are relatively infrequent. As discussed above, variations in KDM expression are much more common [53]. However, still, there are a few exceptions. KDM6A (UTX) is often truncated or mutated in many cancers [54, 55]. Additionally, KDM5A fused to the nuclear pore component Nup98 was described in AML, and this fusion caused leukemia when expressed in mice [56, 57]. Thus, changes in the expression level or gene integrity of KDM genes all contribute to carcinogenesis.

Given the therapeutic potential of targeting KDMs, academia and industry are pursuing heavily histone lysine demethylase inhibitors (HDMi) as novel chemotherapy agents. When employing HDMi as sensitizing agents, there are two potential mechanisms to sensitize tumor cells to chemotherapy: (1) converting quiescent cells that used to be insensitive to drugs to an activated form by manipulating the expressions of drug-tolerant genes or drug-targeted genes, and (2) reestablishing the balance of methylation levels that control the expressions of oncogenes or tumor suppressors. Over the years, plenty of compounds targeting KDM have been developed. The rest of this chapter will provide an update on a few HDMi being investigated in overcoming chemoresistance (Table 2).

TABLE 2 Aberrant KDMs associated with chemotherapy resistance in cancers.

Demethylase	Pathways involved	Cancer type	Chemoresistance
KDM1A	E-cadherin - cell differentiation	AML	ATRA
KDM1A	MYC-cell proliferation	AML	ATRA
KDM1A	Prickle1, APC-β-catenin-Lgr5$^+$ CSCs proliferation	HCC	Cisplatin, sorafenib
KDM2B	DNA repair	Glioblastoma	Lomustine, etoposide
KDM3A	Sox2, Nanog, Bcl-2-CSCs proliferation	Ovarian cancer	Cisplatin
KDM4B	Plk1-cell proliferation	Prostate cancer	Cisplatin
KDM5A	G2 checkpoint	Glioblastoma	TMZ
KDM5A		NSCLC	Cisplatin
KDM5B	CD34$^-$ CSCs proliferation	Melanoma	BRAFi
KDM5B		Neuroblastoma	Cisplatin
KDM6A/6B	PRKCA, MCL1-RAF/ERK/MAPK, Bcl-2-CSCs proliferation	Osteosarcoma	Cisplatin
KDM6B		Lymphoma	R-CHOP
KDM6A/6B	NOTCH2-stemness genes	Colorectal cancer	Oxaliplatin

AML, acute myeloid leukemia; *HCC*, hepatocellular carcinoma; *NSCLC*, nonsmall cell lung cancer; *CSCs*, cancer stem cells; *ATRA*, all-trans retinoic acid; *TMZ*, temozolomide.

KDM1/LSD inhibitors as chemo-sensitizing agents

Because the catalytic domain of KDM1/LSD enzymes is homologous to that of MAO, some MAO inhibitors (MAOi), such as tranylcypromine (TCP), are among the first generation of LSD inhibitors. TCP is an irreversible inhibitor which bonds to the LSD cofactor FAD [58], and it has been effective in combination therapy approaches [35, 59], notably promoting AML differentiation, thus sensitizing non-APL AML to ATRA [35]. However, due to lack of specificity for LSD [60], the use of TCP as an LSD inhibitor is limited by the common side effects from anti-MAO activities [61, 62]. Huge efforts have been devoted to improve TCP specificity for LSD, resulting in a large group of compounds with substitutions at the *para* position and/or at the primary amine of TCP (a few examples are demonstrated in Fig. 1). These TCP derivatives effectively inhibit cancer cell growth in vitro or in vivo [63–68]. A systematic comparison of current LSD inhibitors will help us better understand design rationales that confer TCP specificity for LSD.

Currently, the majority of drugs developed against LSD are still structurally based on MAOi. Besides TCP-based irreversible drugs, reversible MAOi scaffolds, such as polyamines and pyrimidines [69, 70], have also been used to generate LSD inhibitors. For example, Bizine (Fig. 1), derived from the irreversible MAOi phenelzine, shows specificity for LSD1 over

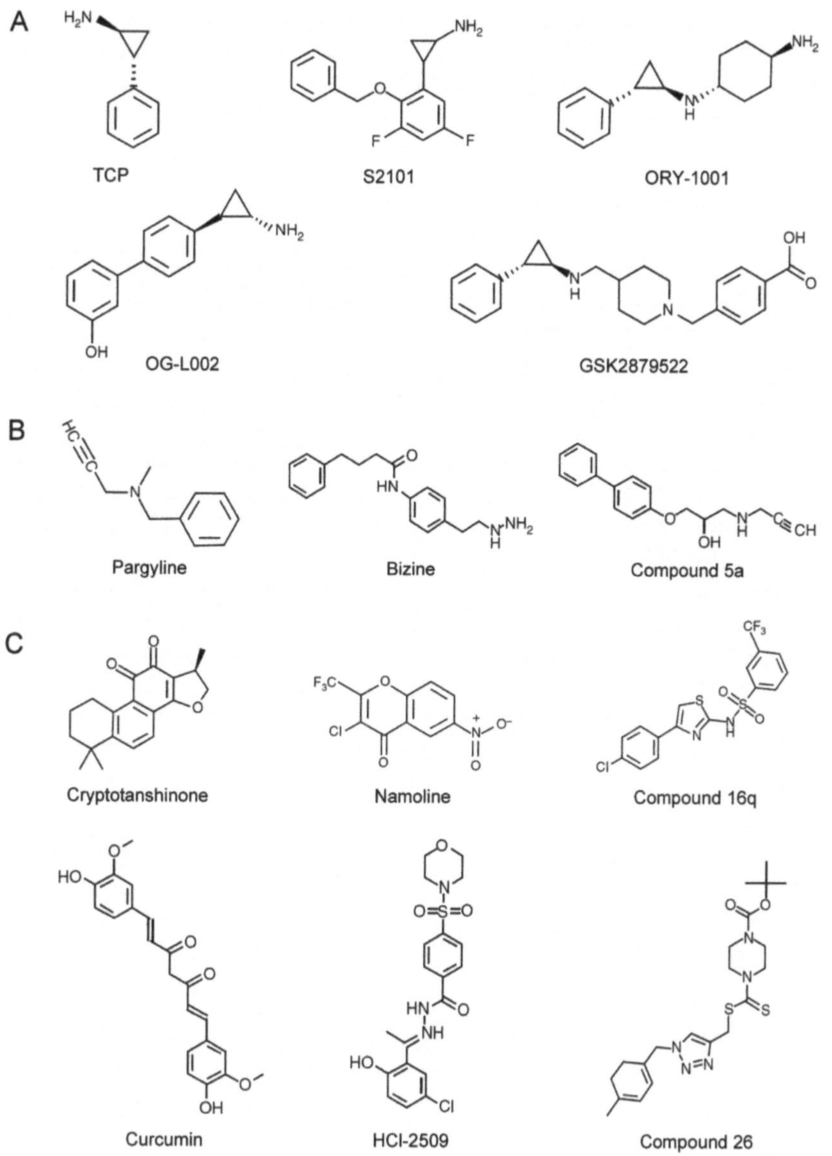

FIG. 1 Chemical structures of selected KDM1 inhibitors. (A) Irreversible KDM1 inhibitor TCP and its derivatives. (B) Irreversible KDM1 inhibitors that are not structurally based on TCP. (C) Reversible KDM1 inhibitors.

MAO [71]. Several research groups were also able to develop LSD1-specific inhibitors not based on any MAOi structure, such as Namoline, cryptotanshinone, compound 16q (aminothiazoles), and the natural compound Curcumin (Fig. 1) [72–75].

Despite the numerous classes of LSD inhibitors being developed, so far only a few (includes TCP, ORY-1001, and GSK2879552) are in clinical trials, and all aim at treating AML or SCLC [20, 76]. The clinical applications of most LSD inhibitors discovered so far remain to be further explored. A more extended review of the structure and specificity of currently available LSD inhibitors can be found elsewhere [16].

JmjC demethylase inhibitors as chemo-sensitizing agents

Given the summarized studies above and additional evidence linking the biological activity of JmjC-containing demethylase to cancers, there is growing interest in developing HDMi for this class of enzymes. Similar to using MAOi for LSD, the initial targeting to JmjC KDMs were performed by inhibitors of 2-OG oxygenases, because of their structural similarity. Major efforts have been devoted to optimizing mimics of the 2-OG cofactor or developing inhibitors that interfere with Fe(II) binding [77, 78]. Two primary scaffolds for 2-OG-dependent enzyme inhibitors are *N*-oxalylglycine (NOG) and *para* 2,4 dicarboxylic acid (2,4-PDCA) (Fig. 2) [79–81]. Based on those scaffolds, several inhibitors have been developed successfully, such as hydroxamate derivatives, *N*-oxalyl amino acid derivatives, pyridine dicarboxylates, and disulfiram [78, 82–85].

Besides NOG and 2,4-PDCA, 8-hydroxyquinoline derivatives have been identified in many large-scale screenings as lead scaffolds for JmjC KDMs. IOX1 (5-carboxy-8-hydroxyquinoline) (Fig. 2) was initially identified as a KDM4 inhibitor [86, 87], and later was derivatized to improve its cell permeability and specificity [88, 89]. Another screening yielded JIB-04 (Fig. 2) as a pan JmjC HDMi [90]. Furthermore, guided by the crystal structure of KDM6A, GSK-J1 and its cell-permeable analog GSK-J4 (Fig. 2) were derived from a lead compound which was reported to be KDM6 specific [91].

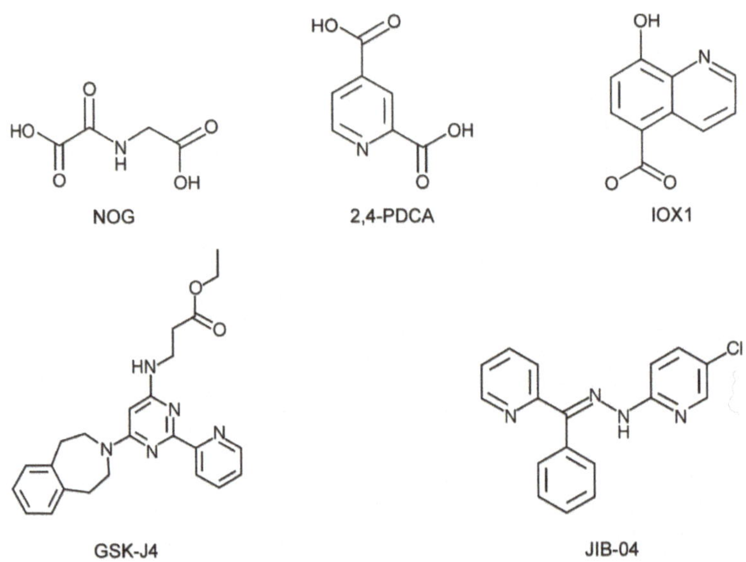

FIG. 2 Chemical structures of selected JmjC-family inhibitors.

Concluding remarks

The success of HDMi in sensitizing many types of cancers in vitro and in vivo shows a great potential for these compounds in clinical use. Although a considerable number of clinical trials currently are at different phases [20, 76, 92], the major challenge is achieving specific inhibition of a defined demethylase, as most HDMi thus far act on related enzymes (other KDMs as well as non-KDM enzymes with similar catalytic mechanisms). Recent advances in compound screening, structure-orientated design, and structure-activity studies will facilitate both the specificity and the efficacy of HDMi.

Another challenge for HDMi in cancer therapy is that for now no HDMi has shown any kind of cell-type or tissue specificity. Given the broad roles of demethylases, a fine balance of histone methylation body-wide is critical for both maintenance of normal cellular homeostasis and control of unrestrained tumor metastasis. Thus, ideally, HDMi for chemosensitization should be able to work only on tumor cells but not normal cells. The current rapid development of targeted drug delivery systems, such as nanomedicine and novel tissue-specific biomaterials, provides promising strategies which would enable HDMi to selectively reach their site of action without reaching the nontarget cells, organs, or tissues [93, 94].

Lastly, it should be noted that dysregulation of KDMs in cancers is not necessarily equal to upregulation of KDMs in cancers, as in some cases KDMs show reduced expression and/or activities in certain types of cancers [95]. For example, KDM5B is considered as a tumor suppressor because it shows downregulation, mutation, or loss in prostate cancer and renal cell carcinoma [96, 97]. Another example is that the loss of LSD1 in breast cancer promotes tumor invasion via upregulating TGF-β [98]. In such a scenario, HDMi might contribute to worse outcomes [95].

Acknowledgment

This work was supported by NSFC grants (U1601227, 82070406) and the Department of Science and Technology of Guangdong (2015B020225006, 2020A1515011158).

References

[1] Bernstein BE, Meissner A, Lander ES. The mammalian epigenome. Cell 2007;128(4):669–81.
[2] Luger K, Mäder AW, Richmond RK, Sargent DF, Richmond TJ. Crystal structure of the nucleosome core particle at 2.8 Å resolution. Nature 1997;389(6648):251–60.
[3] Jenuwein T. Translating the histone code. Science 2001;293(5532):1074–80.
[4] Gelato KA, Fischle W. Role of histone modifications in defining chromatin structure and function. Biol Chem 2008;389(4):353–63.
[5] Barski A, Cuddapah S, Cui K, Roh T-Y, Schones DE, Wang Z, et al. High-resolution profiling of histone methylations in the human genome. Cell 2007;129(4):823–37.
[6] Kouzarides T. Chromatin modifications and their function. Cell 2007;128(4):693–705.
[7] Lee JS, Smith E, Shilatifard A. The language of histone crosstalk. Cell 2010;142(5):682–5.
[8] Taverna SD, Li H, Ruthenburg AJ, Allis CD, Patel DJ. How chromatin-binding modules interpret histone modifications: lessons from professional pocket pickers. Nat Struct Mol Biol 2007;14(11):1025–40.
[9] Chang B, Chen Y, Zhao Y, Bruick RK. JMJD6 is a histone arginine demethylase. Science 2007;318(5849):444–7.

[10] Walport LJ, Hopkinson RJ, Chowdhury R, Schiller R, Ge W, Kawamura A, et al. Arginine demethylation is catalysed by a subset of JmjC histone lysine demethylases. Nat Commun 2016;7(1):11974.
[11] Shi Y, Lan F, Matson C, Mulligan P, Whetstine JR, Cole PA, et al. Histone demethylation mediated by the nuclear amine oxidase homolog LSD1. Cell 2004;119(7):941–53.
[12] Tsukada Y, Fang J, Erdjument-Bromage H, Warren ME, Borchers CH, Tempst P, et al. Histone demethylation by a family of JmjC domain-containing proteins. Nature 2006;439(7078):811–6.
[13] Karytinos A, Forneris F, Profumo A, Ciossani G, Battaglioli E, Binda C, et al. A novel mammalian flavin-dependent histone demethylase. J Biol Chem 2009;284(26):17775–82.
[14] Tian X, Fang J. Current perspectives on histone demethylases. Acta Biochim Biophys Sin Shanghai 2007; 39(2):81–8.
[15] Dong C, Zhang H, Xu C, Arrowsmith CH, Min J. Structure and function of dioxygenases in histone demethylation and DNA/RNA demethylation. IUCrJ 2014;1(Pt 6):540–9.
[16] Jambhekar A, Anastas JN, Shi Y. Histone lysine demethylase inhibitors. Cold Spring Harb Perspect Med 2017; 7(1):a026484.
[17] Batlle E, Clevers H. Cancer stem cells revisited. Nat Med 2017;23(10):1124–34.
[18] Steinbichler TB, Dudás J, Skvortsov S, Ganswindt U, Riechelmann H, Skvortsova I-I. Therapy resistance mediated by cancer stem cells. Semin Cancer Biol 2018;53:156–67.
[19] Toh TB, Lim JJ, Chow EK-H. Epigenetics in cancer stem cells. Mol Cancer 2017;16(1):29.
[20] Morera L, Lübbert M, Jung M. Targeting histone methyltransferases and demethylases in clinical trials for cancer therapy. Clin Epigenetics 2016;8:57.
[21] McCabe MT, Mohammad HP, Barbash O, Kruger RG. Targeting histone methylation in cancer. Cancer J 2017; 23(5):292–301.
[22] McGrath J, Trojer P. Targeting histone lysine methylation in cancer. Pharmacol Ther 2015;150:1–22.
[23] Cancer Genome Atlas Research Network. Comprehensive molecular profiling of lung adenocarcinoma. Nature 2014;511(7511):543–50.
[24] Bailey MH, Tokheim C, Porta-Pardo E, Sengupta S, Bertrand D, Weerasinghe A, et al. Comprehensive characterization of cancer driver genes and mutations. Cell 2018;173(2):371–85.
[25] Cooper LA, Demicco EG, Saltz JH, Powell RT, Rao A, Lazar AJ. PanCancer insights from the cancer genome atlas: the pathologist's perspective. J Pathol 2018;244(5):512–24.
[26] Gillette MA, Satpathy S, Cao S, Dhanasekaran SM, Vasaikar SV, Krug K, et al. Proteogenomic characterization reveals therapeutic vulnerabilities in lung adenocarcinoma. Cell 2020;182(1):200–25.
[27] Chen YJ, Roumeliotis TI, Chang YH, Chen CT, Han CL, Lin MH, et al. Proteogenomics of non-smoking lung cancer in East Asia delineates molecular signatures of pathogenesis and progression. Cell 2020;182(1):226–44.
[28] Xu JY, Zhang C, Wang X, Zhai L, Ma Y, Mao Y, et al. Integrative proteomic characterization of human lung adenocarcinoma. Cell 2020;182(1):245–61.
[29] Hayami S, Kelly JD, Cho HS, Yoshimatsu M, Unoki M, Tsunoda T, et al. Overexpression of LSD1 contributes to human carcinogenesis through chromatin regulation in various cancers. Int J Cancer 2011;128(3):574–86.
[30] Schildhaus H-U, Riegel R, Hartmann W, Steiner S, Wardelmann E, Merkelbach-Bruse S, et al. Lysine-specific demethylase 1 is highly expressed in solitary fibrous tumors, synovial sarcomas, rhabdomyosarcomas, desmoplastic small round cell tumors, and malignant peripheral nerve sheath tumors. Hum Pathol 2011; 42(11):1667–75.
[31] Harris WJ, Huang X, Lynch JT, Spencer GJ, Hitchin JR, Li Y, et al. The histone demethylase KDM1A sustains the oncogenic potential of MLL-AF9 leukemia stem cells. Cancer Cell 2012;21(4):473–87.
[32] Mohammad HP, Smitheman KN, Kamat CD, Soong D, Federowicz KE, Van Aller GS, et al. A DNA hypomethylation signature predicts antitumor activity of LSD1 inhibitors in SCLC. Cancer Cell 2015; 28(1):57–69.
[33] Coombs CC, Tavakkoli M, Tallman MS. Acute promyelocytic leukemia: where did we start, where are we now, and the future. Blood Cancer J 2015;5, e304.
[34] Murray-Stewart T, Woster PM, Casero RA. The re-expression of the epigenetically silenced e-cadherin gene by a polyamine analogue lysine-specific demethylase-1 (LSD1) inhibitor in human acute myeloid leukemia cell lines. Amino Acids 2014;46(3):585–94.
[35] Schenk T, Chen WC, Göllner S, Howell L, Jin L, Hebestreit K, et al. Inhibition of the LSD1 (KDM1A) demethylase reactivates the all-trans-retinoic acid differentiation pathway in acute myeloid leukemia. Nat Med 2012; 18(4):605–11.

[36] Lei ZJ, Wang J, Xiao HL, Guo Y, Wang T, Li Q, et al. Lysine-specific demethylase 1 promotes the stemness and chemoresistance of Lgr5(+) liver cancer initiating cells by suppressing negative regulators of β-catenin signaling. Oncogene 2015;34(24):3188–98.

[37] Sharma SV, Lee DY, Li B, Quinlan MP, Takahashi F, Maheswaran S, et al. A chromatin-mediated reversible drug-tolerant state in cancer cell subpopulations. Cell 2010;141(1):69–80.

[38] Pippa S, Manniron C, Licursi V, Bombardi L, Colotti G, Cundari E, et al. Small molecule inhibitors of KDM5 histone demethylases increase the radiosensitivity of breast cancer cells overexpressing JARID1B. Molecules 2019;24(9):1739.

[39] Kuo YT, Liu YL, Adebayo BO, Shih PH, Lee WH, Wang LS, et al. JARID1B expression plays a critical role in chemoresistance and stem cell-like phenotype of neuroblastoma cells. PLoS One 2015;10(5), e0125343.

[40] Banelli B, Carra E, Barbieri F, Würth R, Parodi F, Pattarozzi A, et al. The histone demethylase KDM5A is a key factor for the resistance to temozolomide in glioblastoma. Cell Cycle 2015;14(21):3418–29.

[41] Liu X, Zhang SM, McGeary MK, Krykbaeva I, Lai L, Jansen DJ, et al. KDM5B promotes drug resistance by regulating melanoma-propagating cell subpopulations. Mol Cancer Ther 2019;18(3):706–17.

[42] Staberg M, Rasmussen RD, Michaelsen SR, Pedersen H, Jensen KE, Villingshøj M, et al. Targeting glioma stem-like cell survival and chemoresistance through inhibition of lysine-specific histone demethylase KDM2B. Mol Oncol 2018;12(3):406–20.

[43] Ramadoss S, Sen S, Ramachandran I, Roy S, Chaudhuri G, Farias-Eisner R. Lysine-specific demethylase KDM3A regulates ovarian cancer stemness and chemoresistance. Oncogene 2017;36(11):1537–45.

[44] Duan L, Rai G, Roggero C, Zhang QJ, Wei Q, Ma SH, et al. KDM4/JMJD2 histone demethylase inhibitors block prostate tumor growth by suppressing the expression of AR and BMYB-regulated genes. Chem Biol 2015; 22(9):1185–96.

[45] He C, Sun J, Liu C, Jiang Y, Hao Y. Elevated H3K27me3 levels sensitize osteosarcoma to cisplatin. Clin Epigenetics 2019;11(1):8.

[46] Mathur R, Sehgal L, Havranek O, Köhrer S, Khashab T, Jain N, et al. Inhibition of demethylase KDM6B sensitizes diffuse large B-cell lymphoma to chemotherapeutic drugs. Haematologica 2017;102(2):373–80.

[47] Wang Q, Chen X, Jiang Y, Liu S, Liu H, Sun X, et al. Elevating H3K27me3 level sensitizes colorectal cancer to oxaliplatin. J Mol Cell Biol 2020;12(2):125–37.

[48] Berry WL, Shin S, Lightfoot SA, Janknecht R. Oncogenic features of the JMJD2A histone demethylase in breast cancer. Int J Oncol 2012;41(5):1701–6.

[49] Yamane K, Tateishi K, Klose RJ, Fang J, Fabrizio LA, Erdjument-Bromage H, et al. PLU-1 is an H3K4 demethylase involved in transcriptional repression and breast cancer cell proliferation. Mol Cell 2007;25(6):801–12.

[50] Wang Q, Wei J, Su P, Gao P. Histone demethylase JARID1C promotes breast cancer metastasis cells via down regulating BRMS1 expression. Biochem Biophys Res Commun 2015;464(2):659–66.

[51] Saydaminova K, Strauss R, Xie M, Bartek J, Richter M, van Rensburg R, et al. Sensitizing ovarian cancer cells to chemotherapy by interfering with pathways that are involved in the formation of cancer stem cells. Cancer Biol Ther 2016;17(10):1079–88.

[52] Dalvi MP, Wang L, Zhong R, Kollipara RK, Park H, Bayo J, et al. Taxane-platin-resistant lung cancers co-develop hypersensitivity to JumonjiC demethylase inhibitors. Cell Rep 2017;19(8):1669–84.

[53] Chen QW, Zhu XY, Li YY, Meng ZQ. Epigenetic regulation and cancer (review). Oncol Rep 2014;31(2):523–32.

[54] van Haaften G, Dalgliesh GL, Davies H, Chen L, Bignell G, Greenman C, et al. Somatic mutations of the histone H3K27 demethylase gene UTX in human cancer. Nat Genet 2009;41(5):521–3.

[55] Dalgliesh GL, Furge K, Greenman C, Chen L, Bignell G, Butler A, et al. Systematic sequencing of renal carcinoma reveals inactivation of histone modifying genes. Nature 2010;463(7279):360–3.

[56] van Zutven LJCM, Onen E, Velthuizen SCJM, van Drunen E, von Bergh ARM, van den Heuvel-Eibrink MM, et al. Identification of NUP98 abnormalities in acute leukemia: JARID1A (12p13) as a new partner gene. Genes Chromosom Cancer 2006;45(5):437–46.

[57] Gough SM, Lee F, Yang F, Walker RL, Zhu YJ, Pineda M, et al. NUP98-PHF23 is a chromatin-modifying oncoprotein that causes a wide array of leukemias sensitive to inhibition of PHD histone reader function. Cancer Discov 2014;4(5):564–77.

[58] Schmidt DMZ, McCafferty DG. Trans-2-Phenylcyclopropylamine is a mechanism-based inactivator of the histone demethylase LSD1. Biochemistry 2007;46(14):4408–16.

[59] Singh MM, Manton CA, Bhat KP, Tsai WW, Aldape K, Barton MC, et al. Inhibition of LSD1 sensitizes glioblastoma cells to histone deacetylase inhibitors. Neuro-Oncology 2011;13(8):894–903.

[60] Lee MG, Wynder C, Schmidt DM, McCafferty DG, Shiekhattar R. Histone H3 lysine 4 demethylation is a target of nonselective antidepressive medications. Chem Biol 2006;13(6):563–7.
[61] Fiedorowicz JG, Swartz KL. The role of monoamine oxidase inhibitors in current psychiatric practice. J Psychiatr Pract 2004;10(4):239–48.
[62] Shulman KI, Fischer HD, Herrmann N, Huo CY, Anderson GM, Rochon PA. Current prescription patterns and safety profile of irreversible monoamine oxidase inhibitors: a population-based cohort study of older adults. J Clin Psychiatry 2009;70(12):1681–6.
[63] Binda C, Valente S, Romanenghi M, Pilotto S, Cirilli R, Karytinos A, et al. Biochemical, structural, and biological evaluation of tranylcypromine derivatives as inhibitors of histone demethylases LSD1 and LSD2. J Am Chem Soc 2010;132(19):6827–33.
[64] Mimasu S, Umezawa N, Sato S, Higuchi T, Umehara T, Yokoyama S. Structurally designed trans-2-phenylcyclopropylamine derivatives potently inhibit histone demethylase LSD1/KDM1. Biochemistry 2010;49(30):6494–503.
[65] Benelkebir H, Hodgkinson C, Duriez PJ, Hayden AL, Bulleid RA, Crabb SJ, et al. Enantioselective synthesis of tranylcypromine analogues as lysine demethylase (LSD1) inhibitors. Bioorg Med Chem 2011;19(12):3709–16.
[66] Pollock JA, Larrea MD, Jasper JS, McDonnell DP, McCafferty DG. Lysine-specific histone demethylase 1 inhibitors control breast cancer proliferation in ERα-dependent and -independent manners. ACS Chem Biol 2012; 7(7):1221–31.
[67] Valente S, Rodriguez V, Mercurio C, Vianello P, Saponara B, Cirilli R, et al. Pure enantiomers of benzoylamino-tranylcypromine: LSD1 inhibition, gene modulation in human leukemia cells and effects on clonogenic potential of murine promyelocytic blasts. Eur J Med Chem 2015;94:163–74.
[68] Wu F, Zhou C, Yao Y, Wei L, Feng Z, Deng L, et al. 3-(Piperidin-4-ylmethoxy)pyridine containing compounds are potent inhibitors of lysine specific demethylase 1. J Med Chem 2016;59(1):253–63.
[69] George T, Kaul CL, Grewal RS, Tahilramani R. Antihypertensive and monoamine oxidase inhibitory activity of some derivatives of 3-formyl-4-oxo-4H-pyrido (1,2-a) pyrimidine. J Med Chem 1971;14(10):913–5.
[70] Bianchi M, Polticelli F, Ascenzi P, Botta M, Federico R, Mariottini P, et al. Inhibition of polyamine and spermine oxidases by polyamine analogues. FEBS J 2006;273(6):1115–23.
[71] Prusevich P, Kalin JH, Ming SA, Basso M, Givens J, Li X, et al. A selective phenelzine analogue inhibitor of histone demethylase LSD1. ACS Chem Biol 2014;9(6):1284–93.
[72] Willmann D, Lim S, Wetzel S, Metzger E, Jandausch A, Wilk W, et al. Impairment of prostate cancer cell growth by a selective and reversible lysine-specific demethylase 1 inhibitor. Int J Cancer 2012;131(11):2704–9.
[73] Wu CY, Hsieh CY, Huang KE, Chang C, Kang HY. Cryptotanshinone down-regulates androgen receptor signaling by modulating lysine-specific demethylase 1 function. Int J Cancer 2012;131(6):1423–34.
[74] Hitchin JR, Blagg J, Burke R, Burns S, Cockerill MJ, Fairweather EE, et al. Development and evaluation of selective, reversible LSD1 inhibitors derived from fragments. Med Chem Commun 2013;4(11):1513.
[75] Abdulla A, Zhao X, Yang F. Natural polyphenols inhibit lysine-specific demethylase-1 in vitro. J Biochem Pharmacol Res 2013;1(1):56–63.
[76] Maes T, Carceller E, Salas J, Ortega A, Buesa C. Advances in the development of histone lysine demethylase inhibitors. Curr Opin Pharmacol 2015;23:52–60.
[77] Chen Z, Zang J, Kappler J, Hong X, Crawford F, Wang Q, et al. Structural basis of the recognition of a methylated histone tail by JMJD2A. Proc Natl Acad Sci U S A 2007;104(26):10818–23.
[78] Sekirnik R, Rose NR, Thalhammer A, Seden PT, Mecinović J, Schofield CJ. Inhibition of the histone lysine demethylase JMJD2A by ejection of structural Zn(II). Chem Commun (Camb) 2009;(42):6376–8.
[79] Tschank G, Raghunath M, Günzler V, Hanauske-Abel HM. Pyridinedicarboxylates, the first mechanism-derived inhibitors for prolyl 4-hydroxylase, selectively suppress cellular hydroxyprolyl biosynthesis. Decrease in interstitial collagen and Clq secretion in cell culture. Biochem J 1987;248(3):625–33.
[80] Baader E, Tschank G, Baringhaus KH, Burghard H, Günzler V. Inhibition of prolyl 4-hydroxylase by oxalyl amino acid derivatives in vitro, in isolated microsomes and in embryonic chicken tissues. Biochem J 1994; 300(Pt 2):525–30.
[81] Elkins JM, Hewitson KS, McNeill LA, Seibel JF, Schlemminger I, Pugh CW, et al. Structure of factor-inhibiting hypoxia-inducible factor (HIF) reveals mechanism of oxidative modification of HIF-1 alpha. J Biol Chem 2003;278(3):1802–6.
[82] Rose NR, Ng SS, Mecinović J, Liénard BMR, Bello SH, Sun Z, et al. Inhibitor scaffolds for 2-oxoglutarate-dependent histone lysine demethylases. J Med Chem 2008;51(22):7053–6.

[83] Hamada S, Kim TD, Suzuki T, Itoh Y, Tsumoto H, Nakagawa H, et al. Synthesis and activity of N-oxalylglycine and its derivatives as Jumonji C-domain-containing histone lysine demethylase inhibitors. Bioorg Med Chem Lett 2009;19(10):2852–5.
[84] Rose NR, Woon ECY, Kingham GL, King ONF, Mecinović J, Clifton IJ, et al. Selective inhibitors of the JMJD2 histone demethylases: combined nondenaturing mass spectrometric screening and crystallographic approaches. J Med Chem 2010;53(4):1810–8.
[85] Itoh Y, Sawada H, Suzuki M, Tojo T, Sasaki R, Hasegawa M, et al. Identification of Jumonji AT-rich interactive domain 1A inhibitors and their effect on cancer cells. ACS Med Chem Lett 2015;6(6):665–70.
[86] King ONF, Li XS, Sakurai M, Kawamura A, Rose NR, Ng SS, et al. Quantitative high-throughput screening identifies 8-hydroxyquinolines as cell-active histone demethylase inhibitors. PLoS One 2010;5(11), e15535.
[87] Hopkinson RJ, Tumber A, Yapp C, Chowdhury R, Aik W, Che KH, et al. 5-Carboxy-8-hydroxyquinoline is a broad spectrum 2-oxoglutarate oxygenase inhibitor which causes iron translocation. Chem Sci 2013; 4(8):3110–7.
[88] Schiller R, Scozzafava G, Tumber A, Wickens JR, Bush JT, Rai G, et al. A cell-permeable ester derivative of the JmjC histone demethylase inhibitor IOX1. ChemMedChem 2014;9(3):566–71.
[89] Feng T, Li D, Wang H, Zhuang J, Liu F, Bao Q, et al. Novel 5-carboxy-8-HQ based histone demethylase JMJD2A inhibitors: introduction of an additional carboxyl group at the C-2 position of quinoline. Eur J Med Chem 2015;105:145–55.
[90] Wang L, Chang J, Varghese D, Dellinger M, Kumar S, Best AM, et al. A small molecule modulates Jumonji histone demethylase activity and selectively inhibits cancer growth. Nat Commun 2013;4:2035.
[91] Kruidenier L, Chung C, Cheng Z, Liddle J, Che K, Joberty G, et al. A selective jumonji H3K27 demethylase inhibitor modulates the proinflammatory macrophage response. Nature 2012;488(7411):404–8.
[92] Fang Y, Liao G, Yu B. LSD1/KDM1A inhibitors in clinical trials: advances and prospects. J Hematol Oncol 2019;12(1):129.
[93] Hare JI, Lammers T, Ashford MB, Puri S, Storm G, Barry ST. Challenges and strategies in anti-cancer nanomedicine development: an industry perspective. Adv Drug Deliv Rev 2017;108:25–38.
[94] Li Z, Xiao C, Yong T, Li Z, Gan L, Yang X. Influence of nanomedicine mechanical properties on tumor targeting delivery. Chem Soc Rev 2020;49(8):2273–90.
[95] Park SY, Park JW, Chun YS. Jumonji histone demethylases as emerging therapeutic targets. Pharmacol Res 2016;105:146–51.
[96] Li N, Dhar SS, Chen TY, Kan PY, Wei Y, Kim JH, et al. JARID1D is a suppressor and prognostic marker of prostate cancer invasion and metastasis. Cancer Res 2016;76(4):831–43.
[97] Arseneault M, Monlong J, Vasudev NS, Laskar RS, Safisamghabadi M, Harnden P, et al. Loss of chromosome Y leads to down regulation of KDM5D and KDM6C epigenetic modifiers in clear cell renal cell carcinoma. Sci Rep 2017;7:44876.
[98] Wang Y, Zhang H, Chen Y, Sun Y, Yang F, Yu W, et al. LSD1 is a subunit of the NuRD complex and targets the metastasis programs in breast cancer. Cell 2009;138(4):660–72.

CHAPTER

5

Targeting histone lysine methyltransferases for drug sensitization

Ao Shen and Xi-Yong Yu

Key Laboratory of Molecular Target & Clinical Pharmacology and the State Key Laboratory of Respiratory Disease, School of Pharmaceutical Sciences & The Fifth Affiliated Hospital, Guangzhou Medical University, Guangzhou, People's Republic of China

Abstract

Histone methylation is part of the most common epigenetic regulations that play important roles in developmental processes and diseases. The methylation status of lysine residues in histone proteins determines the active and silent states of surrounding genes by modulating the chromatin architecture. In cancers, via direct or indirect regulation of tumor suppressors and oncogenes, aberrant activities of histone lysine methyltransferases facilitate oncogenic effects including tumor proliferation and invasion, chemotherapy resistance, and disease relapse. Consequently, discovering and developing compounds targeting histone lysine methyltransferases have grown to be a very attractive research area over the past two decades. In this chapter, we generally reviewed the connection between histone methylation and cancers and discussed the molecular functions underlying chemotherapy resistance and the clinical significance of targeting histone lysine methyltransferases in drug sensitization.

Abbreviations

AML	acute myeloid leukemia
CRC	colorectal cancer
CSC	cancer stem cell
KMT	histone lysine methyltransferase
MLL	mixed-lineage leukemia
NSCLC	non-small cell lung cancer
OS	osteosarcoma
PARP	poly-ADP-ribose polymerase
PRC2	polycomb repressive complex 2
PRMT	histone arginine methyltransferase

https://doi.org/10.1016/B978-0-12-823684-0.00003-7

SAM	S-adenosylmethionine
SCLC	small cell lung cancer
SET	Su(var)3-9, E(z), Trx
T-ALL	T cell acute lymphoblastic leukemia

Conflict of interest

No potential conflicts of interest were disclosed.

An overview of histone methylation

Gene expressions are essential for cellular homeostasis and are controlled not only by DNA sequence, transcription factors, and ubiquitin-dependent proteolytic machinery, but also by epigenetic regulations [1]. Within the eukaryotic cell nucleus, DNA is packaged together with histones to form a complex known as chromatin. Chromatin can be contained in a relatively loose and transcriptionally active form called euchromatin, or in a densely compacted and transcriptionally repressive form called heterochromatin [2]. Different epigenetic regulations influence the chromatin state, thus controlling accessibility to DNA, genome stability, and other DNA metabolic processes [3]. These regulations mainly go through DNA methylation [4], nucleosome remodeling [5], noncoding RNA [6], and histone posttranslational modifications [7]. This chapter is focused on histone modifications, specifically histone methylation.

The nucleosome, which is the basic repeating unit of chromatin, is constituted by an octameric core of histones (containing two copies each of H2A, H2B, H3, and H4) wrapped by DNA. The nucleosome is approximately globular, except for the amino termini of histones protruding from it, on which many residues are subjected to various types of posttranslational modification. Acetylation, methylation, phosphorylation, ubiquitination, and sumoylation are among the most common histone modifications, which are essential for maintaining genomic integrity, and play an important role in determining gene expression [3, 8]. Additional types of histone modifications have been documented in recent years including, but not limited to, formylation, propionylation, serotonylation, malonylation, and succinylation, although their impacts on gene expression are much less studied [3, 8].

Histone methylation occurs at various sites in histone proteins, primarily occurring on lysines (K) and arginines (R) on the H3 and H4 tails. Methyl groups (me) can be sequentially added and result in monomethylated, dimethylated, or trimethylated lysine, or monomethylated, symmetrical, or asymmetrical dimethylated arginine. A conventional code refers to a description of the status of histone methylation. The substrate is indicated by the histone subunit and an IUPAC code for the amino acid and number of the amino acid being modified. The degree of methylation is also noted by the addition of "me#," with the numbers being 1, 2, and 3 for monomethylated, dimethylated, and trimethylated substrates, respectively. For example, H3R17me2a means histone H3 with asymmetric dimethylated arginine, which is the 17th amino acid of H3.

For histone lysine methylation, the most studied sites are K4, K9, K27, K36, K79 on histone H3 and K20 on histone H4 (Table 1). Other methylated lysines are also found in H1, H2A, H2B, and additional sites in H3 and H4. While methylations at some sites are mainly associated with gene activation (like H3K4me and H4R3me2a) or repression (like H3K9me2 and H4R3me2s) [9], more often the outcome of histone methylation is governed by diverse regulators, as well as other modifications on the same or other histones. Consistently, even at

TABLE 1 Systematic nomenclature of histone lysine methyltransferases and their histone substrates.

Subfamily	Symbols	Aliases	Primary target site	Secondary target site
KMT1	SUV39H1	KMT1A, MG44	H3K9	
	SUV39H2	KMT1B		
	EHMT2	KMT1C, G9A, GAT8		
	EHMT1	KMT1D, GLP		
	SETDB1	KMT1E, ESET		
	SETDB2	KMT1F, CLLD8		
KMT2	KMT2A	MLL1, CXXC7	H3K4	
	KMT2B	MLL2, MLL4, CXXC10		
	KMT2C	MLL3		
	KMT2D	MLL2, MLL4		
	KMT2E	MLL5		
	SETD1A	KMT2F, SET1A, EPEDD		
	SETD1B	KMT2G, SET1B		
	ASH1L	KMT2H, ASH1		
KMT3	SETD2	KMT3A, SET2, HIF-1	H3K36	
	NSD1	KMT3B, SOTOS, STO		
	NSD2	KMT3F/G, MMSET		
	NSD3	KMT3F/G		
	SMYD2	KMT3C, ZMYND14	H3K4	H3K36
	SMYD1	KMT3D, ZMYND18	H3K4	
	SMYD3	KMT3E, ZMYND1	H3K4	H4K5
KMT4	DOT1L	KMT4, DOT1	H3K79	
KMT5	KMT5A	SET8, SETD8	H4K20	
	KMT5B	SUV420H1, CGI85		
	KMT5C	SUV420H2		
KMT6	EZH2	KMT6A, ENX1	H3K27	
	EZH1	KMT6B		
KMT7	SETD7	KMT7, SET7, SET9	H3K4	
KMT8	PRDM2	KMT8A, RIZ1	H3K4	H3K9
	PRDM9	KMT8B, MEISETZ		

a single site, histone methylation can either activate or repress gene expression, which depends on the degree of methylation. In recent years, the complexity of histone methylation and demethylation has expanded further, with the identification of more positive and negative regulators [10]. Cellular homeostasis thus requires a precise balance of histone methylation and demethylation, and arginine and lysine methylation together create an enormously diversified array of histone methylation patterns [2, 11]. Such meticulous methylation marks are mainly orchestrated by three types of epigenetic regulators, namely the "writers" (methyltransferases), the "erasers" (demethylases), and the "readers" (transcriptional factors that specifically recognize methylation patterns) [10, 12]. Consequently, changes of these methylation patterns are implicated in both normal development and diseases like cancer. A more extended review of histone modifications can be found in Chapter 4.

Histone lysine methyltransferases

Histone methyltransferases are the "writer" enzymes that catalyze the transfer of methyl groups to either lysine or arginine residues of histones. Histone arginine methyltransferases (PRMTs) are made up of a small family of about nine proteins in humans which can be divided into three different types, based on whether they promote monomethylation, asymmetrical or symmetrical dimethylation of arginine [13]. For histone lysine methyltransferases (KMTs), there are two types: SET (Su(var)3–9, E(z), Trx) domain-containing KMTs and DOT1-like KMTs. Since first reported 20 years ago for SUV39H1 (KMT1A), the discovery and characterization of KMTs have progressed rapidly [14]. Now the human KMT family contains more than 50 members and further can be subdivided into eight subfamilies (from KMT1 to KMT8), of which all but one are SET domain-containing KMTs (Table 1). Each subfamily of KMTs acts on different substrates and plays different roles in cellular function.

The SET domain was first recognized as a conserved sequence in three Drosophila proteins: Suppressor of variegation 3–9 (Su(var)3–9), Enhancer of zeste (E(z)), and Trithorax (Trx). The SET domain, the pre- and post-SET domains that flank the SET domain are all engaged in methyltransferase activity. The SET domain itself contains a catalytic core for KMT enzymatic activity, and the interplay between the pre-SET and SET regions helps to determine lysine site specificity, while the post-SET domain can be absent in some KMTs, including the well-studied EZH1 and EZH2 [15, 16]. The SET domain-containing KMTs generally function in multiprotein complexes. The accessory proteins in the complex contribute to the selectivity and activity of the complex. Methyltransferases within each KMT subfamily have similar sequence motifs surrounding the SET domain and higher similarity in the SET domain. The different subfamilies of SET-containing KMTs highlight the specific and dynamic regulation of methylation at different histones and various lysine residues.

Unlike other KMT subfamilies, DOT1-like KMT (DOT1L or KMT4) is the only non-SET domain-containing KMT and it specifically methylates the nucleosomal histone H3. While the SET domain-containing KMTs mainly modify the lysines in the tail region of histones, DOT1L methylates K79 of histone H3 in the core domain, and is the only enzyme known to do so [15]. The catalytic domain of DOT1L is structurally more akin to PRMT [17, 18]. The N terminal of DOT1L contains the active catalytic domain, and the C-terminal of DOT1 is critical for nucleosome binding and the substrate specificity [17]. Unlike histone demethylases that use either flavin adenine dinucleotide (FAD) or 2-oxoglutarate (2-OG) as cofactors, S-adenosylmethionine (SAM) serves as the only cofactor and methyl donor for all currently known histone methyltransferases, despite the distinct structural feature and substrate affinity of the SET and non-SET domain-containing KMTs [19].

Dysregulation of KMTs in cancers

Epigenetics regulate how and when each gene is turned on and off, impacting literally all biological pathways for the maintenance of cellular homeostasis. Emerging evidence indicates that epigenetic regulators are often compromised in a wide range of cancer types, particularly in cancer stem cells (CSCs) of those cancers that drive cancer recurrence [20]. It is now well accepted that aberrant histone methylation, in particular hypermethylation, influences the pathobiology of many types of cancer by promotion of tumor proliferation and invasion, chemotherapy resistance and disease relapse [14, 16, 21, 22]. Among those oncogenic effects, chemotherapy resistance is a major challenge for cancer treatment. Chemotherapy often is not able to eliminate CSCs and, more importantly, drug-treated tumors usually harbor a higher frequency of drug-resistant CSCs [23, 24]. The hypermethylation is mainly caused by either a defect of histone demethylases and/or hyperactivity of histone methyltransferases. Instead of trying to restore defective demethylase activity, using small molecules as histone lysine methyltransferase inhibitors to reduce the methylation level of targets is more feasible in potential clinical applications. As a consequence, KMTs are valuable targets for therapeutic intervention.

Recent advances in the cancer genome atlas and cancer proteogenomic projects provide us a comprehensive understanding of both oncogenic and tumor suppressive functions of KMTs [25–30]. For instance, in one project, 299 genes were found to drive tumor progression. Of these, eight genes are associated with histone methylation and demethylation. Among them, five KMTs including SETD2 (KMT3A) and MLL1–4 (KMT2A-D) were designated as cancer driver genes [26]. In many other cases, KMTs are either somatically altered, or their expressions are dysregulated early in tumor progression, suggesting that these enzymes are critical in control of unrestrained tumor metastasis [25–30]. Although the exact mechanisms are yet to be determined in many cases, the oncogenic effects of KMTs are attributed to direct or indirect regulation of either tumor suppressors or oncogenes, or both. Studies since then have identified KMTs as one of the most frequently dysregulated enzymes in many blood or solid tumors. Some of the most studied examples include EZH2, G9a, and DOT1L, which will be briefly reviewed below.

Targeting EZH2 for drug sensitization

EZH2 and its paralog EZH1 are the catalytic components of the polycomb repressive complex 2 (PRC2) complex involved in global transcriptional repression by methylating H3K27me3, mainly targeting tumor suppressor genes. EZH2 is commonly overexpressed in various types of solid cancers such as ovarian cancer, pancreatic cancer, lung cancer, and it also has "gain-of-function" activating mutations in several types of lymphoma and leukemia. Extensive studies have uncovered an important role for EZH2 in cancer progression and chemotherapy resistance, therefore suggesting EZH2 as an effective therapeutic target.

In pancreatic cancer, EZH2 has more nuclear accumulation in poorly differentiated pancreatic adenocarcinomas and reduces the expression of p27 (Kip1), a tumor suppressor gene. Genetic depletion of EZH2 results in upregulation of p27 and decreases pancreatic cancer cell proliferation. Moreover, EZH2 depletion sensitizes pancreatic cancer cells to doxorubicin and gemcitabine, suggesting that the combination of EZH2 inhibitors and standard chemotherapy could be used to enhance chemo-sensitivity for pancreatic cancer [31]. In another report, Curcumin, a phenolic compound extracted from turmeric, can sensitize pancreatic CSCs to various chemotherapeutic agents through the inhibition of the EZH2-lncPVT1-cMyc axis [32].

For lung cancers, cisplatin treatment induces high expression of HOXB13 in non-small cell lung cancer (NSCLC), in turn upregulates ABCG1/EZH2/Slug expression and distant metastasis. These regulatory changes account for a novel mechanism for cisplatin resistance and metastasis after chemotherapy [33]. In small cell lung cancer (SCLC) chemoresistant models, EZH2 induces local chromatin condensation and suppression of SLFN11, a factor implicated in DNA damage repair deficiency, which is associated with cisplatin and etoposide resistance [34]. Consequently, inclusion of an EZH2 inhibitor with standard chemotherapies could prevent acquired cisplatin resistance and augment chemotherapeutic efficacy for lung cancers.

For Acute myeloid leukemia (AML), PRC2 maintains the stemness of AML CSCs which is responsible for AML relapse after conventional chemotherapy. Inactivation of EZH1/2 by a novel EZH1/2 dual inhibitor can eradicate quiescent CSCs to cure AML. Mechanistically, this is achieved by reversion of PRC2-mediated suppression of Cyclin D, inducing cell cycle progression of quiescent CSCs and differentiation [35]. For prostate cancer, EZH2 depletion or EZH2 inhibitor GSK126 overcomes enzalutamide resistance in both cultured cells and xenograft tumors. Mechanistically, EZH2 directly binds to the promoter of prostate-specific antigen and inhibits its expression in enzalutamide-resistant cancer cells [36]. In glioblastoma, EZH2-NOTCH1 signaling is the key pathway to maintain glioblastoma CSCs, and melatonin inhibits glioblastoma CSCs through suppression of the EZH2-NOTCH1 axis. Surprisingly, instead of transcriptionally silencing NOTCH1 by H3K27me3, EZH2 regulates NOTCH1 expression by directly binding to the NOTCH1 promoter [37].

EZH2 might also work together with other types of epigenetic regulators to modulate chemoresistance. In two recent reports regarding cisplatin resistance in gastric cancer, EZH2 can be recruited by lncRNA HOXD-AS1 or lncRNA PCAT-1, to silence PDCD4 or PTEN, respectively, and confers cisplatin resistance [38, 39].

It should be noted that dysregulation of EZH2 in cancers is not necessarily equal to upregulation of EZH2 in cancers, as in many cases EZH2 shows reduced expression and/or activities in certain types of cancers. For example, poly-ADP-ribose polymerase inhibitors (PARPi) have been widely used in treatment of BRCA1/2-deficient tumors; however, stabilization of stalled DNA replication forks is a recently identified PARPi-resistance mechanism. EZH2 localizes at stalled forks, mediating recruitment of the MUS81 nuclease which can promote genomic stability. Accordingly, low EZH2 expression predicts chemoresistance and poor outcome in patients with BRCA2-mutated tumors. Moreover, inhibition of EZH2 in a murine Brca2$^{-/-}$ breast tumor model is linked to acquired PARPi resistance [40]. In another case, after cisplatin treatment, cisplatin-resistant osteosarcoma (OS) specimens have a lower H3K27me3 level than cisplatin-sensitive specimens. In vitro inhibition of EZH2 in OS cells decreased H3K27me3 levels and led to cisplatin resistance. Accordingly, inhibition of KDM6 can reverse OS cisplatin resistance both in vitro and in vivo [41]. Mechanistically, decreased H3K27me3 level on stemness-related genes such as PRKCA and MCL1 results in activation of these genes, ultimately activating the RAF/ERK/MAPK pathway and antiapoptotic Bcl-2, rendering OS resistant to cisplatin [41].

In a most recent report on colorectal cancer, a low H3K27me3 level predicts a poor outcome upon chemotherapeutic treatment. Mechanistically, oxaliplatin stimulation induces NOTCH2 expression, which promotes the expression of stemness-related genes and results in oxaliplatin resistance. This is achieved by downregulation of the H3K27me3 level on the NOTCH2 transcription initiation site due to overexpression of KDM6. Elevation of H3K27me3 through the KDM6 inhibitor GSK-J4 treatment significantly enhances oxaliplatin-induced apoptosis.

Conversely, inhibiting H3K27me3 by the EZH2 inhibitor EPZ-6438 remarkably decreases apoptotic cancer cells [42].

In T cell acute lymphoblastic leukemia (T-ALL) clinical specimens, loss-of-function mutations of PRC2 core components (EZH2, EED, or SUZ12) are linked to cytotoxic chemotherapy resistance. In T-ALL cells, PRC2 depletion induced resistance to multiple chemotherapeutics. Mechanistically, EZH2 inhibition causes upregulation of transcription factor CRIP2 and downstream upregulation of the mitochondrial chaperone TRAP1, which leads to chemoresistance [43]. Consistently, EZH2 levels correlated with poor AML prognosis. Suppression of EZH2 expression induces acquired resistance to tyrosine kinase inhibitors and several cytotoxic drugs in AML cell lines and primary cells in vitro and in vivo. Mechanistically, low EZH2 expression results in upregulation of HOXB7 and HOXA9, and knockdown of HOX genes in resistant cells improves drug sensitivity [44]. These data suggest restoration of EZH2, rather than repression of EZH2, as a viable approach to overcome drug resistance in certain types of cancers.

G9a, DOT1L, and other KMTs in drug sensitization

The EZH1/2 (KMT6) subfamily is one of the most studied histone-modifying enzymes in cancer research. There are also other dysregulated KMT subfamilies contributing to oncogenesis. DOT1L (KMT4), the only non-SET domain-containing KMT, also plays a pivotal role in regulating the malignant ovarian cancer. By recruiting DOT1L, C/EBPβ, an independent prognostic factor in ovarian cancer patients, can maintain an open chromatin state by H3K79 methylation of multiple drug-resistance genes, thereby raising the platinum resistance of tumor cells [45].

The Fas-FasL signaling plays a critical role in cancer immune surveillance by host T cells, but metastatic human colorectal cancer (CRC) often skips the immune system via silencing Fas expression. SUV39H1 (KMT1A) expression is significantly upregulated in CRC. SUV39H1-selective inhibitor F5446 decreases H3K9me3 at the Fas gene promoter, thus increasing CRC cell sensitivity to FasL-induced apoptosis. F5446 also alters the expression of genes implicated in DNA replication and cell cycle in the metastatic CRC cells. Consequently, F5446 increases CRC sensitivity to both 5-fluorouracil and FasL-induced apoptosis in vitro, and inhibits tumor growth in vivo [46]. Consistently, another report found that using verticillin A in CRC, when it acted as a KMT1 inhibitor, decreased H3K9me3 in the FAS promoter and restored Fas expression. Furthermore, verticillin A treatment resensitizes CRC to multiple drugs, including DR5 agonist drozitumab and 5-fluorouracil in vitro and in vivo [47]. The success of using verticillin A for pancreas ductal adenocarcinoma, which is highly resistant to chemotherapy, radiotherapy, and immunotherapy, has also been reported. Pancreas ductal adenocarcinoma uses H3K4me3 to activate Bcl-x, FLIP, and Mcl-1, and H3K9me3 to silence Bak, Bax, and Bim to acquire an apoptosis-resistant phenotype. Therefore, selective inhibition of H3K4me3 and H3K9me3 by verticillin A is an effective approach to overcome gemcitabine resistance [48].

Similarly, SUV39H2 (KMT1B), another member of the KMT1 subfamily, is highly expressed in malignant glioma. SUV39H2 knockdown inhibits cell proliferation and stemness and promotes the chemo-sensitivity of glioma cells in vitro and in vivo. This is achieved by repressing HHIP expression and hedgehog signaling [49]. G9a and GLP (G9a-like protein) also belong to the KMT1 subfamily and form a G9a/GLP heterodimeric complex predominantly. G9a/GLP is implicated in chemoresistance in many cancers, such as resistance to PARPi in ovarian carcinoma. Disruption of G9a/GLP resensitizes ovarian carcinoma to PARPi through DNA damage and cell cycle dysregulation [50].

For the KMT2 subfamily, the mixed-lineage leukemia genes (MLL) are the most studied ones due to their fundamental roles in MLL-rearranged leukemias [51, 52]. Mechanistically, MLL1 is found to regulate MDR1 (multidrug resistance) transcription through H3K4me. The latter confers tumor drug resistance by actively effluxing a number of antitumor agents [53]. SETD2 (KMT3A), a well-known tumor suppressor, is often mutated in MLL-rearranged leukemias. SETD2 mutations confer resistance to cytarabine-based chemotherapy in AML partly through altered cell cycle checkpoints. Mechanistically, SETD2-mutant leukemic cells showed downregulated signaling related to cell cycle progression, S, and G2/M checkpoint regulation [54].

Additionally, despite the fact of being a therapeutic target in many cancers, G9a is found to be the suppressor of aggressive lung tumor propagating cells. G9a inhibition accelerates disease progression. Instead, demethylase inhibition impairs lung adenocarcinoma progression. In this scenario, inhibition of G9a is dangerous in certain cancer contexts, and targeting the histone demethylases seems to be a more suitable approach for lung cancer treatment [55]. Therefore, understanding the cellular context and specific tumor populations is essential when targeting KMTs in cancers.

Conclusion and outlook

This chapter mainly highlights KMTs as a promising pharmacological target for the treatment of cancers, specifically for drug sensitization, and suggests that reestablishing the balance of methylation levels in cancer could be an effective therapeutic strategy. Besides KMTs, arginine methyltransferases PRMTs are also involved in cancer drug resistance. For example, a novel PRMT5 inhibitor DW14800 targeting liver cancer stem cells inhibits the malignancy of hepatocellular carcinoma by promoting the transcription of HNF4α [56]. Another example is PRMT1, which is recruited via DNA-PK to chromatin and is responsible for cisplatin resistance in ovarian cancer [57]. More extended reviews on the roles of PRMTs in chemotherapy resistance can be found elsewhere [13, 58].

Besides directly targeting histone lysine methyltransferases, working through noncoding RNA, mostly through lncRNA or miRNA, as an indirect histone methyltransferase inhibitor has not been discussed in this chapter. This is an alternative way to small molecule compounds and has been actively pursued [59–66]. Moreover, the modulation of aberrant histone methylation also could be addressed through action on the "readers." This strategy was successful for histone acetylation, but at present few drugs of the histone methylation readers have been reported.

The dynamics of histone methylation and demethylation in different cancers have been reviewed in Chapter 4 and this chapter. Taken together, it is clear that hypermethylation of some portions of the genome and hypomethylation of others are very frequent in cancers. Thus, another challenge for targeting histone methylation in cancer therapy is the improvement of cell-type or tissue specificity. A fine balance of histone methylation body-wide is critical for both maintenance of normal cellular homeostasis and control of unrestrained tumor metastasis. Thus, ideally, histone methylation drugs for chemo-sensitization should be able to target only tumor cells but not normal cells. The current rapid development of targeted drug delivery systems, such as nanomedicine and novel tissue-specific biomaterials, has great translational potential for improving the cell or tissue selectivity [67, 68].

Acknowledgments

This work was supported by NSFC grants (U1601227, 82070406) and the Department of Science and Technology of Guangdong (2015B020225006, 2020A1515011158).

References

[1] Bernstein BE, Meissner A, Lander ES. The mammalian epigenome. Cell 2007;128(4):669–81.
[2] Kouzarides T. Chromatin modifications and their function. Cell 2007;128(4):693–705.
[3] Jenuwein T. Translating the histone code. Science 2001;293(5532):1074–80.
[4] Koh KP, Rao A. DNA methylation and methylcytosine oxidation in cell fate decisions. Curr Opin Cell Biol 2013;25(2):152–61.
[5] Hargreaves DC, Crabtree GR. ATP-dependent chromatin remodeling: genetics, genomics and mechanisms. Cell Res 2011;21(3):396–420.
[6] Wilusz JE, Sunwoo H, Spector DL. Long noncoding RNAs: functional surprises from the RNA world. Genes Dev 2009;23(13):1494–504.
[7] Rothbart SB, Strahl BD. Interpreting the language of histone and DNA modifications. Biochim Biophys Acta 2014;1839(8):627–43.
[8] Gelato KA, Fischle W. Role of histone modifications in defining chromatin structure and function. Biol Chem 2008;389(4):353–63.
[9] Barski A, Cuddapah S, Cui K, Roh TY, Schones DE, Wang Z, et al. High-resolution profiling of histone methylations in the human genome. Cell 2007;129(4):823–37.
[10] Hyun K, Jeon J, Park K, Kim J. Writing, erasing and reading histone lysine methylations. Exp Mol Med 2017; 49(4), e324.
[11] Lee JS, Smith E, Shilatifard A. The language of histone crosstalk. Cell 2010;142(5):682–5.
[12] Taverna SD, Li H, Ruthenburg AJ, Allis CD, Patel DJ. How chromatin-binding modules interpret histone modifications: lessons from professional pocket pickers. Nat Struct Mol Biol 2007;14(11):1025–40.
[13] Jarrold J, Davies CC. PRMTs and arginine methylation: cancer's best-kept secret? Trends Mol Med 2019; 25(11):993–1009.
[14] Husmann D, Gozani O. Histone lysine methyltransferases in biology and disease. Nat Struct Mol Biol 2019; 26(10):880–9.
[15] Dillon SC, Zhang X, Trievel RC, Cheng X. The SET-domain protein superfamily: protein lysine methyltransferases. Genome Biol 2005;6(8):227.
[16] Morera L, Lübbert M, Jung M. Targeting histone methyltransferases and demethylases in clinical trials for cancer therapy. Clin Epigenetics 2016;8:57.
[17] Min J, Feng Q, Li Z, Zhang Y, Xu RM. Structure of the catalytic domain of human DOT1L, a non-SET domain nucleosomal histone methyltransferase. Cell 2003;112(5):711–23.
[18] Nguyen AT, Zhang Y. The diverse functions of Dot1 and H3K79 methylation. Genes Dev 2011;25(13):1345–58.
[19] Serefidou M, Venkatasubramani AV, Imhof A. The impact of one carbon metabolism on histone methylation. Front Genet 2019;10:764.
[20] Toh TB, Lim JJ, Chow EK-H. Epigenetics in cancer stem cells. Mol Cancer 2017;16(1):29.
[21] McGrath J, Trojer P. Targeting histone lysine methylation in cancer. Pharmacol Ther 2015;150:1–22.
[22] McCabe MT, Mohammad HP, Barbash O, Kruger RG. Targeting histone methylation in cancer. Cancer J 2017; 23(5):292–301.
[23] Batlle E, Clevers H. Cancer stem cells revisited. Nat Med 2017;23(10):1124–34.

[24] Steinbichler TB, Dudás J, Skvortsov S, Ganswindt U, Riechelmann H, Skvortsova I-I. Therapy resistance mediated by cancer stem cells. Semin Cancer Biol 2018;53:156–67.

[25] Cancer Genome Atlas Research Network. Comprehensive molecular profiling of lung adenocarcinoma. Nature 2014;511(7511):543–50.

[26] Bailey MH, Tokheim C, Porta-Pardo E, Sengupta S, Bertrand D, Weerasinghe A, et al. Comprehensive characterization of cancer driver genes and mutations. Cell 2018;173(2):371–85.

[27] Cooper LA, Demicco EG, Saltz JH, Powell RT, Rao A, Lazar AJ. PanCancer insights from the cancer genome atlas: the pathologist's perspective. J Pathol 2018;244(5):512–24.

[28] Gillette MA, Satpathy S, Cao S, Dhanasekaran SM, Vasaikar SV, Krug K, et al. Proteogenomic characterization reveals therapeutic vulnerabilities in lung adenocarcinoma. Cell 2020;182(1):200–25.

[29] Chen YJ, Roumeliotis TI, Chang YH, Chen CT, Han CL, Lin MH, et al. Proteogenomics of non-smoking lung cancer in East Asia delineates molecular signatures of pathogenesis and progression. Cell 2020;182(1):226–44.

[30] Xu JY, Zhang C, Wang X, Zhai L, Ma Y, Mao Y, et al. Integrative proteomic characterization of human lung adenocarcinoma. Cell 2020;182(1):245–61.

[31] Ougolkov AV, Bilim VN, Billadeau DD. Regulation of pancreatic tumor cell proliferation and chemoresistance by the histone methyltransferase enhancer of zeste homologue 2. Clin Cancer Res 2008;14(21):6790–6.

[32] Yoshida K, Toden S, Ravindranathan P, Han H, Goel A. Curcumin sensitizes pancreatic cancer cells to gemcitabine by attenuating PRC2 subunit EZH2, and the lncRNA PVT1 expression. Carcinogenesis 2017; 38(10):1036–46.

[33] Zhan J, Wang P, Li S, Song J, He H, Wang Y, et al. HOXB13 networking with ABCG1/EZH2/slug mediates metastasis and confers resistance to cisplatin in lung adenocarcinoma patients. Theranostics 2019;9(7):2084–99.

[34] Gardner EE, Lok BH, Schneeberger VE, Desmeules P, Miles LA, Arnold PK, et al. Chemosensitive relapse in small cell lung cancer proceeds through an EZH2-SLFN11 axis. Cancer Cell 2017;31(2):286–99.

[35] Fujita S, Honma D, Adachi N, Araki K, Takamatsu E, Katsumoto T, et al. Dual inhibition of EZH1/2 breaks the quiescence of leukemia stem cells in acute myeloid leukemia. Leukemia 2018;32(4):855–64.

[36] Bai Y, Zhang Z, Cheng L, Wang R, Chen X, Kong Y, et al. Inhibition of enhancer of zeste homolog 2 (EZH2) overcomes enzalutamide resistance in castration-resistant prostate cancer. J Biol Chem 2019;294(25):9911–23.

[37] Zheng X, Pang B, Gu G, Gao T, Zhang R, Pang Q, et al. Melatonin inhibits glioblastoma stem-like cells through suppression of EZH2-NOTCH1 signaling axis. Int J Biol Sci 2017;13(2):245–53.

[38] Ye Y, Yang S, Han Y, Sun J, Xv L, Wu L, et al. HOXD-AS1 confers cisplatin resistance in gastric cancer through epigenetically silencing PDCD4 via recruiting EZH2. Open Biol 2019;9(9):190068.

[39] Li H, Ma X, Yang D, Suo Z, Dai R, Liu C. PCAT-1 contributes to cisplatin resistance in gastric cancer through epigenetically silencing PTEN via recruiting EZH2. J Cell Biochem 2020;121(2):1353–61.

[40] Rondinelli B, Gogola E, Yücel H, Duarte AA, van de Ven M, van der Sluijs R, et al. EZH2 promotes degradation of stalled replication forks by recruiting MUS81 through histone H3 trimethylation. Nat Cell Biol 2017;19(11): 1371–8.

[41] He C, Sun J, Liu C, Jiang Y, Hao Y. Elevated H3K27me3 levels sensitize osteosarcoma to cisplatin. Clin Epigenetics 2019;11(1):8.

[42] Wang Q, Chen X, Jiang Y, Liu S, Liu H, Sun X, et al. Elevating H3K27me3 level sensitizes colorectal cancer to oxaliplatin. J Mol Cell Biol 2020;12(2):125–37.

[43] Ariës IM, Bodaar K, Karim SA, Chonghaile TN, Hinze L, Burns MA, et al. PRC2 loss induces chemoresistance by repressing apoptosis in T cell acute lymphoblastic leukemia. J Exp Med 2018;215(12):3094–114.

[44] Göllner S, Oellerich T, Agrawal-Singh S, Schenk T, Klein H-U, Rohde C, et al. Loss of the histone methyltransferase EZH2 induces resistance to multiple drugs in acute myeloid leukemia. Nat Med 2017; 23(1):69–78.

[45] Liu D, Zhang XX, Li MC, Cao CH, Wan DY, Xi BX, et al. C/EBPβ enhances platinum resistance of ovarian cancer cells by reprogramming H3K79 methylation. Nat Commun 2018;9(1):1739.

[46] Lu C, Klement JD, Yang D, Albers T, Lebedyeva IO, Waller JL, et al. SUV39H1 regulates human colon carcinoma apoptosis and cell cycle to promote tumor growth. Cancer Lett 2020;476:87–96.

[47] Paschall AV, Yang D, Lu C, Choi J-H, Li X, Liu F, et al. H3K9 trimethylation silences Fas expression to confer colon carcinoma immune escape and 5-fluorouracil chemoresistance. J Immunol 2015;195(4):1868–82.

[48] Lu C, Yang D, Sabbatini ME, Colby AH, Grinstaff MW, Oberlies NH, et al. Contrasting roles of H3K4me3 and H3K9me3 in regulation of apoptosis and gemcitabine resistance in human pancreatic cancer cells. BMC Cancer 2018;18(1):149.

[49] Wang R, Cheng L, Yang X, Chen X, Miao Y, Qiu Y, et al. Histone methyltransferase SUV39H2 regulates cell growth and chemosensitivity in glioma via regulation of hedgehog signaling. Cancer Cell Int 2019;19:269.

[50] Watson ZL, Yamamoto TM, McMellen A, Kim H, Hughes CJ, Wheeler LJ, et al. Histone methyltransferases EHMT1 and EHMT2 (GLP/G9A) maintain PARP inhibitor resistance in high-grade serous ovarian carcinoma. Clin Epigenetics 2019;11(1):165.

[51] Rao RC, Dou Y. Hijacked in cancer: the KMT2 (MLL) family of methyltransferases. Nat Rev Cancer 2015;15 (6):334–46.

[52] Winters AC, Bernt KM. MLL-rearranged leukemias-an update on science and clinical approaches. Front Pediatr 2017;5:4.

[53] Huo H, Magro PG, Pietsch EC, Patel BB, Scotto KW. Histone methyltransferase MLL1 regulates MDR1 transcription and chemoresistance. Cancer Res 2010;70(21):8726–35.

[54] Dong Y, Zhao X, Feng X, Zhou Y, Yan X, Zhang Y, et al. SETD2 mutations confer chemoresistance in acute myeloid leukemia partly through altered cell cycle checkpoints. Leukemia 2019;33(11):2585–98.

[55] Rowbotham SP, Li F, Dost AFM, Louie SM, Marsh BP, Pessina P, et al. H3K9 methyltransferases and demethylases control lung tumor-propagating cells and lung cancer progression. Nat Commun 2018;9(1):4559.

[56] Zheng BN, Ding CH, Chen SJ, Zhu K, Shao J, Feng J, et al. Targeting PRMT5 activity inhibits the malignancy of hepatocellular carcinoma by promoting the transcription of HNF4α. Theranostics 2019;9(9):2606–17.

[57] Musiani D, Giambruno R, Massignani E, Ippolito MR, Maniaci M, Jammula S, et al. PRMT1 is recruited via DNA-PK to chromatin where it sustains the senescence-associated secretory phenotype in response to cisplatin. Cell Rep 2020;30(4):1208–22.

[58] Yang Y, Bedford MT. Protein arginine methyltransferases and cancer. Nat Rev Cancer 2013;13(1):37–50.

[59] Bourguignon LYW, Wong G, Shiina M. Up-regulation of histone methyltransferase, DOT1L, by matrix hyaluronan promotes microRNA-10 expression leading to tumor cell invasion and chemoresistance in cancer stem cells from head and neck squamous cell carcinoma. J Biol Chem 2016;291(20):10571–85.

[60] Gu P, Chen X, Xie R, Han J, Xie W, Wang B, et al. lncRNA HOXD-AS1 regulates proliferation and chemoresistance of castration-resistant prostate cancer via recruiting WDR5. Mol Ther 2017;25(8):1959–73.

[61] Li P, Zhang X, Wang H, Wang L, Liu T, Du L, et al. MALAT1 is associated with poor response to oxaliplatin-based chemotherapy in colorectal cancer patients and promotes chemoresistance through EZH2. Mol Cancer Ther 2017;16(4):739–51.

[62] Niu Y, Ma F, Huang W, Fang S, Li M, Wei T, et al. Long non-coding RNA TUG1 is involved in cell growth and chemoresistance of small cell lung cancer by regulating LIMK2b via EZH2. Mol Cancer 2017;16(1):5.

[63] Chen X, Xie R, Gu P, Huang M, Han J, Dong W, et al. Long noncoding RNA LBCS inhibits self-renewal and chemoresistance of bladder cancer stem cells through epigenetic silencing of SOX2. Clin Cancer Res 2019; 25(4):1389–403.

[64] Lv L, Li Q, Chen S, Zhang X, Tao X, Tang X, et al. miR-133b suppresses colorectal cancer cell stemness and chemoresistance by targeting methyltransferase DOT1L. Exp Cell Res 2019;385(1):111597.

[65] Li Q, Song W, Wang J. TUG1 confers Adriamycin resistance in acute myeloid leukemia by epigenetically suppressing miR-34a expression via EZH2. Biomed Pharmacother 2019;109:1793–801.

[66] Zhang Y, Tang B, Song J, Yu S, Li Y, Su H, et al. Lnc-PDZD7 contributes to stemness properties and chemosensitivity in hepatocellular carcinoma through EZH2-mediated ATOH8 transcriptional repression. J Exp Clin Cancer Res 2019;38(1):92.

[67] Hare JI, Lammers T, Ashford MB, Puri S, Storm G, Barry ST. Challenges and strategies in anti-cancer nanomedicine development: an industry perspective. Adv Drug Deliv Rev 2017;108:25–38.

[68] Li Z, Xiao C, Yong T, Li Z, Gan L, Yang X. Influence of nanomedicine mechanical properties on tumor targeting delivery. Chem Soc Rev 2020;49(8):2273–90.

CHAPTER

6

Ribonucleotide base-modifying enzymes and diseases

Masamitsu Konno[a], *Ayumu Asai*[b], *Takaaki Hirotsu*[a,c], *and Hideshi Ishii*[a]

[a]Osaka University Graduate School of Medicine, Center of Medical Innovation and Translational Research, Osaka, Japan [b]Artificial Intelligence Research Center, The Institute of Scientific and Industrial Research, Osaka University, Osaka, Japan [c]Hirotsu Bio Science Inc., Tokyo, Japan

Abstract

The epitranscriptome is a posttranscriptional base modification of RNAs. More than 140 types of RNA base modification have been elucidated so far. The base modifications control RNA functions by contributing to the construction of RNA three-dimensional structures. In recent years, it has emerged that several RNA base-modifying enzymes are involved in the onset, progression, and malignant transformation of diseases through the involvement of epitranscriptomic mechanisms. This review article discusses the relationship between epitranscriptomics and disease so that knowledge of epitranscriptome can be applied for use in diagnostic tools and therapeutic approaches.

Abbreviations

5-AZA	5-azacytidine
5mC	5-methylcytosine
ac4C	N4-acetylcytidine
AKT	AKT serine/threonine kinase
ALKBH8	alkylated DNA repair protein alkB homolog 8
AVIL	advillin
CYP27B1	cytochrome P450, family 27, subfamily B, polypeptide 1
DNMT2	DNA (cytosine-5-)-methyltransferase 2
EEF1AKMT3	EEF1A lysine methyltransferase 3
FASTKD2	FAST kinase domain-containing protein 2, mitochondrial
FTO	FTO alpha-ketoglutarate-dependent dioxygenase
GWAS	genome-wide association studies
HCC	hepatocellular carcinoma
HIV	human immunodeficiency virus

Epigenetic Regulation in Overcoming Chemoresistance
https://doi.org/10.1016/B978-0-12-823684-0.00004-9

IL20RA	interleukin 20 receptor subunit alpha
m6A	N6-methyladenosinne
m7G	N7-methylguanine
METTL	methyltransferase-like protein
MLASA	mitochondrial myopathy, lactic acidosis, sideroblastic anemia
MRM	S-adenosyl-L-methionine-dependent 2′-O-ribose methyltransferase
mRNA	messenger RNA
MS	multiple sclerosis
NAT10	*N*-acetyltransferase 10
NDUFA4L2	NADH dehydrogenase [ubiquinone] 1 alpha subcomplex subunit 4-like 2
NSUN2	NOP2/Sun RNA methyltransferase 2
NXN	nucleoredoxin
PTEN	phosphatase and tensin homolog deleted from chromosome 10
PUS1	pseudouridine synthase 1
RNA	ribonucleotide
rRNA	ribosomal RNA
snoRNA	small nucleolar RNA
SNP	single-nucleotide polymorphism
snRNA	small nuclear RNA
TRMT	tRNA methyltransferase
tRNA	transfer RNA
TSFM	Ts translation elongation factor, mitochondrial
TSPAN31	tetraspanin 31

Conflict of interest

No potential conflicts of interest were disclosed.

Introduction

The role of epitranscriptomics (modification of ribonucleotide, RNA) as a component of disease has been revealed as a result of next-generation sequencing and the development of other related technologies used for nucleotide sequencing and modification [1]. Epitranscriptomics is controlled by modification writes, erases, and reads, and methylation is mediated by enzymatic reactions [2]. More than 140 modified bases and enzymes that promote base modifications have been identified to date [3]. The epitranscriptomic mechanism is involved in the development of various diseases and malignant changes in humans [1].

Regarding modifications of each RNA, the modifications have been identified in messenger RNA (mRNA), ribosomal RNA (rRNA), transfer RNA (tRNA), small nuclear RNA (snRNA), small nucleolar RNA (snoRNA), and other noncoding RNAs in various species, including humans [4, 5]. Although most RNA base modifications occur in tRNA and rRNA, many changes also occur in other RNA species, such as mRNA, snRNA, and snoRNA. In fact, the chemical diversity of RNA modifications facilitates multiple roles in various biological processes from the developmental stage through to stem cell differentiation [6]. Noncoding RNAs also undergo modification, and their role in controlling gene expression has been investigated [7]. Although some enzymatic reactions are not fully understood, various chemical modifications and their reactions at each RNA base (adenine, guanine, uridine, and cytosine) have been investigated and found to be associated with mild diseases. This review explains the relationship between these RNA base modifications and diseases.

Guanine-modification enzymes

METTL1

Methyltransferase-like protein (METTL) 1 is an enzyme for tRNA, mRNA, and mitochondrial RNA that transforms guanine in RNA into N7-methylguanine (m7G) [8–10] (Fig. 1). METTL1 is upregulated in hepatocellular carcinoma (HCC) and exhibits oncogenic activity via the phosphatase and tensin homolog deleted from the chromosome 10 (PTEN)/AKT serine/threonine kinase (AKT) signaling pathway. High levels of METTL1 expression were found to correlate with larger tumor size, higher serum alpha-fetoprotein levels, tumor vascular invasion, and poor prognosis in two independent cohorts that included 892 patients with HCC [11]. These data suggest METTL1 as a promising prognostic biomarker and that targeting the METTL1/PTEN axis has therapeutic potential for HCC intervention [11]. METTL1 is also involved in the anticancer drug susceptibility of colon cancer, and the overexpression of METTL1 was sensitized by cisplatin-resistant colon cancer cells by modulating the miR-149-3p/S100A4/p53 axis [12]. METTL1 may also help enhance the efficacy of anticancer drug treatment of colorectal cancer. Furthermore, METTL1 is involved in patients with multiple sclerosis (MS) [13–15]. In an attempt to identify functional variants in the genomic regions around chromosome 12q13.3–12q14.1, which is associated with MS, the single-nucleotide polymorphism (SNP) rs6581155 was shown to tag 18 variants within a region where nine genes were mapped, which was sufficient to model associations. This SNP revealed a complete linkage disequilibrium with other polymorphisms associated with expression levels of EEF1A lysine methyltransferase 3 (EEF1AKMT3), advillin (AVIL), Ts translation elongation factor, mitochondrial (TSFM), tetraspanin 31 (TSPAN31), cytochrome P450, family 27, subfamily B, polypeptide 1 (CYP27B1), and METTL1 in different expression trait loci studies [13]. Recently, a search for genetic risk factors for MS has been conducted using genome-wide association studies (GWAS). As a result, six polymorphisms, including METTL1, have been identified as statistically significant [14], and METTL1 and TSFM

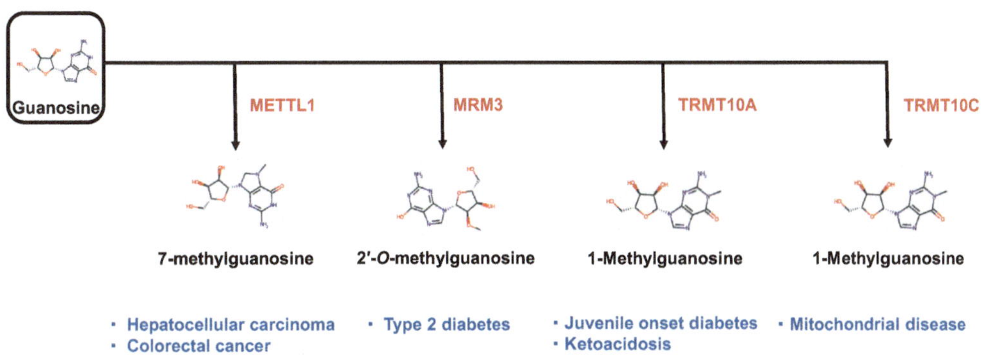

FIG. 1 Guanine-modification enzymes and related diseases. METTL1 is an enzyme that catalyzes guanine to N7-methylguanine (m7G) and is associated with hepatocellular cancer (HCC) and colorectal cancer. MRM3 is an enzyme that catalyzes guanine to 2′-O-methylguanosine and is associated with Type 2 diabetes. TRMT10A is an enzyme that catalyzes guanine to N1-methylguanosine and is associated with juvenile-onset diabetes and ketoacidosis. TRMT10C is another enzyme that catalyzes guanine to N1-methylguanosine, but it is associated with mitochondrial disease.

expression were suggested as involved in MS pathogenesis [15]. Although the detailed mechanism of METTL1 involvement in the onset of MS remains to be elucidated, it is a potential target for MS treatment.

MRM3

S-adenosyl-L-methionine-dependent 2′-O-ribose methyltransferase (MRM3) is a guanine-modification enzyme [16] (Fig. 1). A recent search of the GWAS database and related literature identified 170 genes or genomic regions associated with type 2 diabetes, including interleukin 20 receptor subunit alpha (IL20RA), MRM3, and nucleoredoxin (NXN), that showed considerable differences between populations [17]. Although preliminary, these analyses may provide a better understanding of population differences in susceptibility to type 2 diabetes and insightful references for future studies.

TRMT10A

tRNA methyltransferase 10A (TRMT10A) is an enzyme that catalyzes the methylation of guanine to N1-methylguanine in tRNA [18, 19] (Fig. 1). TRMT10A is involved in juvenile-onset diabetes, microcephaly with short stature, and intellectual disability caused by a nonsense mutation resulting in a stop at position 127 of TRMT10A [20]. Additionally, another group investigated the clinical spectrum presented by a 17-year-old woman with a deletion of homozygous contiguous genes, including TRMT10A. She showed stunting and microcephaly from infancy and a slow, inconsistent progression during adolescence. At the same time, her gonadotropin levels fluctuated between low and high levels, compatible with gonadal failure [21]. However, unlike previous reports, the patient had ketoacidosis at the time of diabetes and the development of autoantibodies against islet cells. Nevertheless, her glycemic control was excellent (glycated hemoglobin, 5.0%–6.2%) [21], suggesting TRMT10A as a therapeutic target for juvenile-onset diabetes and primary human microcephaly.

TRMT10C

tRNA methyltransferase 10C (TRMT10C) is an N1-methyltransferase involved in mitochondrial tRNA maturation [22] (Fig. 1). TRMT10C is also involved in mitochondrial disease, and sequencing identified TRMT10C mutations in patients with lactic acidosis, hypotension, dysphagia, and congenital deafness [23]. Therefore, TRMT10C could be important for RNA processing for correct mitochondrial function.

Cytosine-modification enzymes

DNMT2

DNA (cytosine-5-)-methyltransferase 2 (DNMT2) methylates cytosine 38 in the anticodon loop of tRNA-Asp to form 5-methylcytosine (5mC) [24] (Fig. 2). Human immunodeficiency virus (HIV) 1 enhances its viability in host cells by utilizing this RNA methylation function of

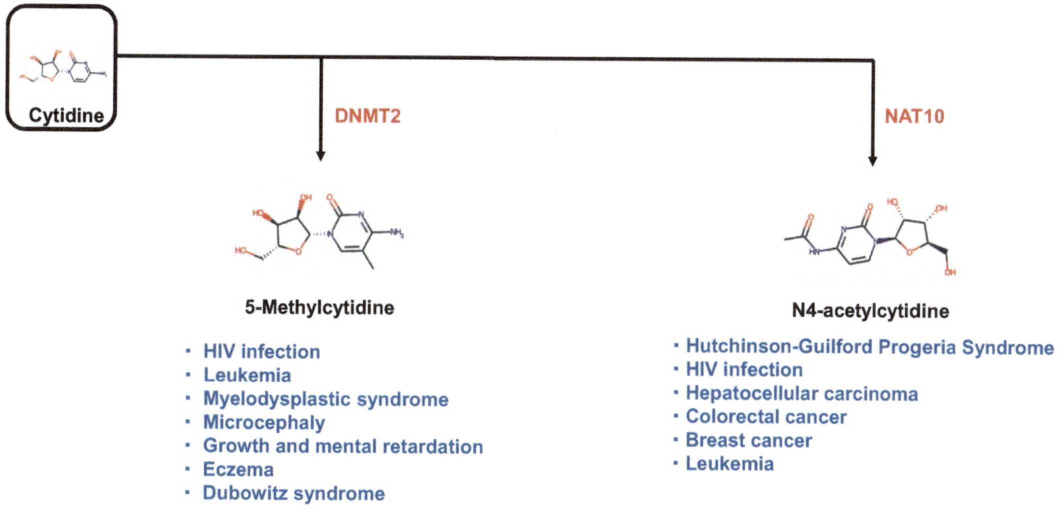

FIG. 2 Cytosine-modification enzymes and related diseases. DNMT2 is an enzyme that catalyzes guanine to 5-methylcytosine and is associated with HIV infection, leukemia, myelodysplastic syndrome, microcephaly, growth, and mental retardation, eczema, and Dubowitz syndrome. NAT10 is an enzyme that catalyzes guanine to N4-acetylcytidine and is associated with HIV infection, HCC, colorectal cancer, breast cancer, and leukemia.

DNMT2 to enhance its genomic stability [25]. Upon HIV infection, DNMT2 relocalizes from the nucleus to the cytoplasm and methylates HIV-1 RNA. This DNMT2-dependent methylation confers posttranscriptional stability on HIV-1 RNA [25]. Furthermore, DNMT2 overexpression increases HIV-1 virus titers and shedding [25]. These results suggest DNMT2 regulation as an effective treatment for HIV. DNMT2 is also involved in acute myelogenous leukemia, and it has been suggested that RNA 5mC is significantly increased in 5-azacytidine (5-AZA)-resistant leukemia cell lines and clinically increased in samples from patients with 5-AZA-resistant myelodysplastic syndrome and acute myeloid leukemia [26]. Thus, DNMT2 may be effective in treating leukemia. Additionally, RNA cytosine C5-methyltransferase, NOP2/Sun RNA methyltransferase 2 (NSUN2) is involved in neurological disorders. A homozygous mutation in NSUN2 that replaces the protein arginine (Gly679Arg) with glycine at position 679 was detected in individuals with autosomal recessive intellectual disability [27]. This amino acid substitution is suggested to interfere with NSUN2 function by preventing the localization of the protein to its site of action in the nucleolus. NSUN2 is also associated with microcephaly, growth, and mental retardation, eczema, and Dubowitz syndrome, which is characterized by distinguishing facial features. Furthermore, homozygous mutations in the canonical splice acceptor of exon 6 led to inexplicable splice donor use, NSUN2 mRNA instability, significantly reduced protein levels, and reduced methylation of NSUN2 target RNA [28].

NAT10

N-acetyltransferase 10 (NAT10) is an RNA cytidine acetyltransferase that catalyzes the formation of N4-acetylcytidine (ac4C) modifications on mRNAs, 18S rRNA, and tRNAs [29–31]

(Fig. 2). Diseases associated with NAT10 include Hutchinson-Gilford progeria syndrome [32], viral infections, and several cancers. Altered nuclei associated with laminopathy, a kind of nuclear envelopathy showing clinically the premature aging disorder as Hutchinson-Guilford progeria syndrome, cause alterations in chromatin organization [32]. Eventually, NAT10 has been identified as a target for Remodelin, which mediates nuclear shape rescue in laminopathic cells via microtubule reorganization [32]. These findings suggest NAT10 as a potential target for treating laminopathy and aging. HIV-1 disrupts NAT10 to increase the expression of viral genes, and the loss of ac4C from viral transcripts due to NAT10 depletion inhibits HIV-1 replication by reducing viral RNA stability [33]. Therefore, NAT10 is expected to be a potential target for antiviral drug development. Moreover, NAT10 is involved in the malignant transformation of various cancers and malignancies, such as glioblastoma [34], HCC [35], colorectal cancer [36, 37], breast cancer [38], and leukemia [39]. In particular, it has been suggested that HCC is involved in distant metastasis via epithelial-mesenchymal transition [40]. Therefore, NAT10 is expected to be a therapeutic target for cancer.

Uridine-modification enzymes

ALKBH8

Alkylated DNA repair protein alkB homolog 8 (ALKBH8) catalyzes uridine to 5-methylcarboxymethyl uridine at the wobble position of the anticodon loop in tRNA via its methyltransferase domain [41, 42] (Fig. 3). ALKBH8 is associated with cancer and is highly expressed in high-grade superficial and deep invasive bladder cancer. This was revealed when ALKBH8 transgenic mice showed an increased rate of bladder tumor mass and invasiveness in an N-butyl-N-(4-hydroxybutyl)-nitrosamine-induced bladder cancer model, suggesting that a high level of ALKBH8 expression is important for bladder cancer growth and progression [43]. Therefore, ALKBH8 may be a drug target in bladder cancer.

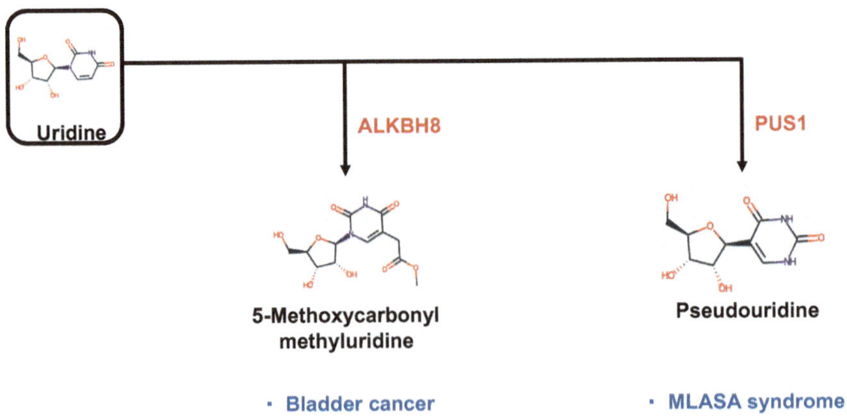

FIG. 3 Uridine-modification enzymes and related diseases. ALKBH8 is an enzyme that catalyzes guanine to 5-methoxycarbonylmethyl uridine and is associated with bladder cancer. PUS1 is an enzyme that catalyzes guanine to pseudouridine and is associated with MLASA syndrome.

PUS1

Pseudouridine synthase 1 (PUS1) converts specific uridines to pseudouridine in several tRNA substrates [44] and is associated with mitochondrial myopathy, lactic acidosis, sideroblastic anemia (MLASA) syndrome [45], a rare autosomal recessive disorder (Fig. 3). Genetic analysis of patients with MLASA syndrome has identified mutations in PUS1 [45–53]. For example, PUS1 mutations c.487delA (I163Lfs*4) and c.884G>A (R295Q) have been identified after childhood in patients with mild cognitive impairment and sideroblastic anemia [54]. Thus, PUS1 mutation could be used to diagnose MLASA syndrome.

Adenine-modification enzymes

TRMT61B

tRNA methyltransferase 61B (TRMT61B) is an enzyme that catalyzes the formation of N1-methyladenine in rRNA and mRNA [55] (Fig. 4A). RNA sequencing of microdissected specimens from subjects with Alzheimer's disease and healthy elderly controls revealed the association of TRMT61B with Alzheimer's disease and confirmed that the expression of mitochondrial-related genes, such as TRMT61B, FAST kinase domain-containing protein 2, mitochondrial (FASTKD2), and NADH dehydrogenase [ubiquinone] 1 alpha subcomplex subunit 4-Like 2 (NDUFA4L2), was increased in Alzheimer's disease patients [56]. A meta-analysis of GWAS revealed that TRMT61B was involved in breast cancer [57]. Therefore, TRMT61B may improve the prediction of breast cancer risk [57].

METTL3

METTL3 forms an N6-methyltransferase complex that methylates adenosine residues at the N6 position of RNA (m6A) [58–60] (Fig. 4A). METTL3 is disrupted by several kinds of viruses, such as Rous sarcoma virus [61], flavivirus [62], Zika virus [63, 64], Kaposi's sarcoma-associated herpesvirus [65, 66], HIV [67], influenza A virus [68], and tobacco mosaic virus [69], to increase the expression of viral genes. Furthermore, METTL3 is associated with neurological diseases, and its association with the severity of Alzheimer's disease [70] and Parkinson's disease [71] has been clarified. Additionally, METTL3 is associated with malignant transformation of various cancers, such as HCC [72–74], nasopharyngeal carcinoma [75, 76], gastric cancer [77, 78], osteosarcoma [79], clear cell renal carcinoma [80], glioblastoma [81], lung cancer [82], ovarian cancer [83], oral squamous cell carcinoma [84, 85], esophageal cancer [86], colorectal cancer [87, 88], and pancreatic cancer [89, 90]. Furthermore, the development of a diagnostic method for cancer using m6A conversion of RNA is in progress [91].

FTO

FTO alpha-ketoglutarate-dependent dioxygenase (FTO) demethylates N6-methyladenosine (m6A) and is the most common internal modification of mRNA [92] (Fig. 4B). FTO is associated with type 2 diabetes because it regulates fat mass, adipogenesis, and body weight, thereby contributing to the regulation of body size and body fat accumulation [93]. GWAS performed by various groups have reported the association of FTO with type 2 diabetes [94–98]. Furthermore, FTO is associated with various cancers and malignancies, such as gastric cancer [99], cervical cancer [100], breast cancer [101, 102], melanoma [103], and leukemia [104].

ALKBH5

Alkylated DNA repair protein alkB homolog 5 (ALKBH5) is another enzyme that demethylates m6A [105] (Fig. 4B). ALKBH5 is associated with various cancers. For example, it suppresses cancer progression in pancreatic cancer and is useful as a diagnostic marker [106, 107]. ALKBH5 is also a useful diagnostic marker in clear cell renal cancer [108]. Furthermore, ALKBH5 is associated with breast cancer [109], ovarian cancer [110], and gastric cancer [111].

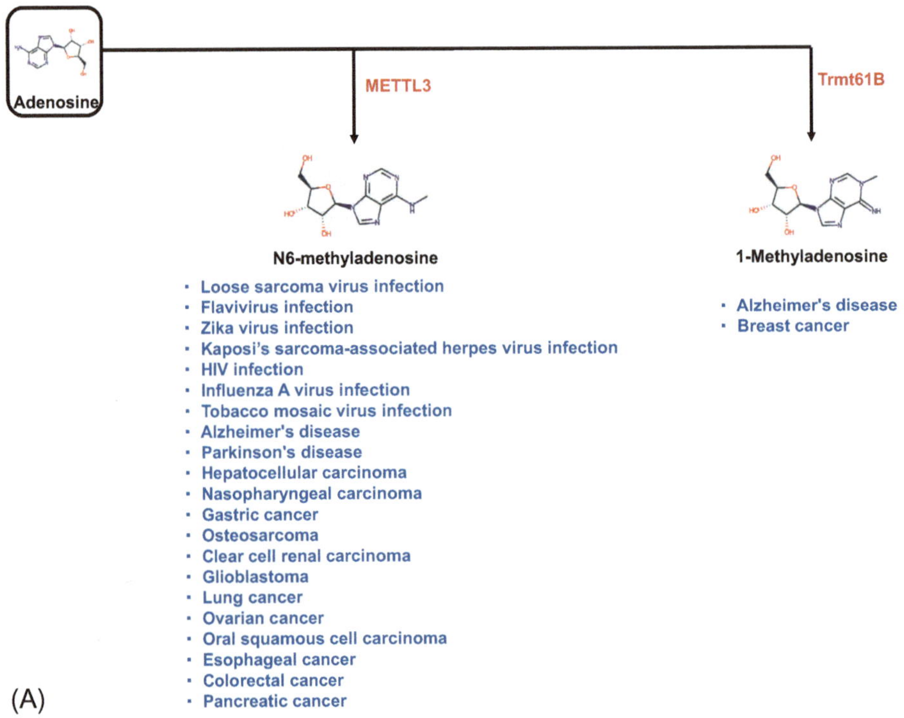

FIG. 4 Adenine-modification enzymes and related diseases. (A) METTL3 is an enzyme that catalyzes guanine to N6-methyladenosine (m6A). It is associated with infections of Rous sarcoma virus, flavivirus, Zika virus, Kaposi's sarcoma-associated herpesvirus, HIV, influenza A virus, and tobacco mosaic virus as well as Alzheimer's disease, Parkinson's disease, HCC, nasopharyngeal carcinoma, gastric cancer, osteosarcoma, clear cell renal carcinoma, glioblastoma, lung cancer, ovarian cancer, oral squamous cell carcinoma, esophageal cancer, colorectal cancer, and pancreatic cancer. TRMT61B is an enzyme that catalyzes guanine to N1-methyladenine and is associated with Alzheimer's disease and breast cancer.

(Continued)

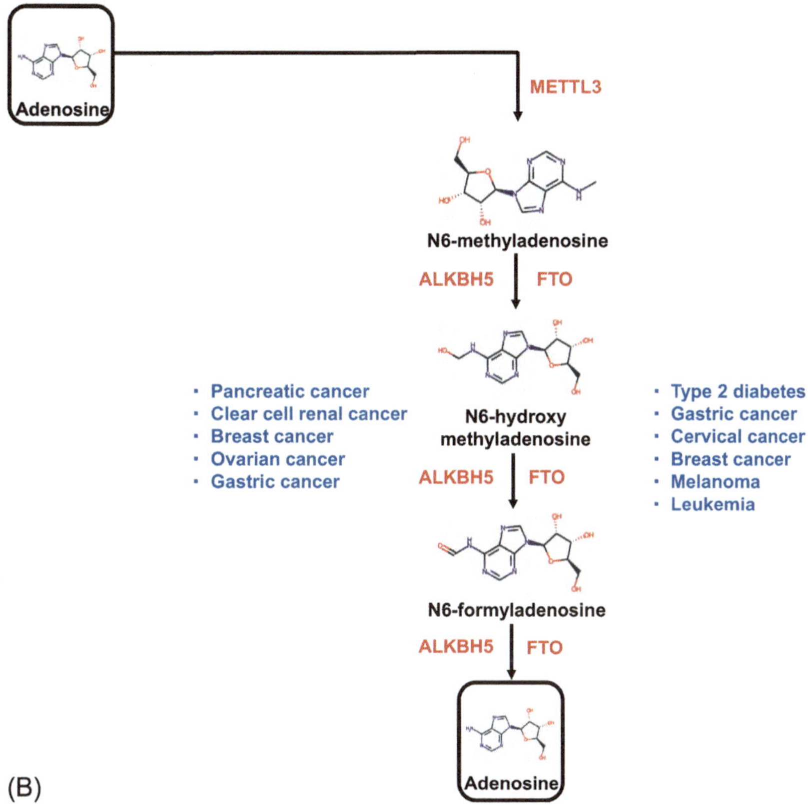

FIG 4, CONT'D (B) ALKBH5 is a demethylation enzyme for m6A and is associated with pancreatic cancer, clear cell renal cancer, breast cancer, ovarian cancer, and gastric cancer. FTO is another demethylation enzyme for m6A and is associated with Type 2 diabetes, gastric cancer, cervical cancer, breast cancer, melanoma, and leukemia.

Conclusion

RNA base-modifying enzymes are associated with various diseases. However, the development of drug discovery and diagnostic methods targeting these enzymes has not yet progressed. It is expected that various diseases will be cured by the development of diagnostic and therapeutic methods targeting the epitranscriptome.

Acknowledgments

We thank the staff in our laboratories. This work was supported in part by Grants-in-Aid for Scientific Research from the Ministry of Education, Culture, Sports, Science, and Technology (17H04282, 17K19698, 18K16356, and 18K16355) and the Japan Agency for Medical Research and Development (16cm0106414h0001 and 17cm0106414h0002). Partial support was received from Princess Takamatsu Cancer Research Fund. Partial institutional endowments were received from Hirotsu Bio Science Inc. (Tokyo, Japan), Kinshu-kai Medical Corporation (Osaka, Japan), Kyowa-kai Medical Corporation (Osaka, Japan), IDEA Consultants Inc. (Tokyo, Japan), and Unitech Co., Ltd. (Chiba, Japan).

References

[1] Jonkhout N, Tran J, Smith MA, Schonrock N, Mattick JS, Novoa EM. The RNA modification landscape in human disease. RNA 2017;23(12):1754–69.

[2] Konno M, Taniguchi M, Ishii H. Significant epitranscriptomes in heterogeneous cancer. Cancer Sci 2019; 110(8):2318–27.

[3] Haruehanroengra P, Zheng YY, Zhou Y, Huang Y, Sheng J. RNA modifications and cancer. RNA Biol 2020;7:1–16.

[4] Frye M, Harada BT, Behm M, He C. RNA modifications modulate gene expression during development. Science 2018;361(6409):1346–9.

[5] Wang X, He C. Dynamic RNA modifications in posttranscriptional regulation. Mol Cell 2014;56(1):5–12.

[6] Ontiveros RJ, Stoute J, Liu KF. The chemical diversity of RNA modifications. Biochem J 2019;476(8):1227–45.

[7] Coker H, Wei G, Brockdorff N. m6A modification of non-coding RNA and the control of mammalian gene expression. Biochim Biophys Acta Gene Regul Mech 2019;1862(3):310–8.

[8] Alexandrov A, Martzen RM, Phizicky ME. Two proteins that form a complex are required for 7-methylguanosine modification of yeast tRNA. RNA 2002;8(10):1253–66.

[9] Zhang L, Liu C, Ma H, Dai Q, Sun H, Luo G, Zhang Z, Zhang L, Hu L, Dong X, He C. Transcriptome-wide mapping of internal N^7-methylguanosine methylome in mammalian mRNA. Mol Cell 2019; 74(6):1304–16.

[10] Pandolfini L, Barbieri I, Bannister JA, Hendrick A, Andrews B, Webster N, Murat P, Mach P, Brandi R, Robson CS, Migliori V, Alendar A, d'Onofrio M, Balasubramanian S, Kouzarides T. METTL1 promotes let-7 microRNA processing via m7G methylation. Mol Cell 2019;74(6):1278–90.

[11] Tian QH, Zhang MF, Zeng JS, Luo RG, Wen Y, Chen J, Gan LG, Xiong JP. METTL1 overexpression is correlated with poor prognosis and promotes hepatocellular carcinoma via PTEN. J Mol Med (Berl) 2019; 97(11):1535–45.

[12] Liu Y, Yang C, Zhao Y, Chi Q, Wang Z, Sun B. Overexpressed methyltransferase-like 1 (METTL1) increased chemosensitivity of colon cancer cells to cisplatin by regulating miR-149-3p/S100A4/p53 axis. Aging (Albany NY) 2019;11(24):12328–44.

[13] Alcina A, Fedetz M, Fernández O, Saiz A, Izquierdo G, Lucas M, Leyva L, García-León JA, Abad-Grau, Mdel M, Alloza I, Antigüedad A, Garcia-Barcina MJ, Vandenbroeck K, Varadé J, de la Hera B, Arroyo R, Comabella M, Montalban X, Petit-Marty N, Navarro A, Otaegui D, Olascoaga J, Blanco Y, Urcelay E, Matesanz F. Identification of a functional variant in the KIF5A-CYP27B1-METTL1-FAM119B locus associated with multiple sclerosis. J Med Genet 2013;50(1):25–33.

[14] Hadjigeorgiou GM, Kountra PM, Koutsis G, Tsimourtou V, Siokas V, Dardioti M, Rikos D, Marogianni C, Aloizou AM, Karadima G, Ralli S, Grigoriadis N, Bogdanos D, Panas M, Dardiotis E. Replication study of GWAS risk loci in Greek multiple sclerosis patients. Neurol Sci 2019;40(2):253–60.

[15] Mo XB, Lei SF, Qian QY, Guo YF, Zhang YH, Zhang H. Integrative analysis revealed potential causal genetic and epigenetic factors for multiple sclerosis. J Neurol 2019;266(11):2699–709.

[16] Lee KW, Bogenhagen DF. Assignment of 2′-O-methyltransferases to modification sites on the mammalian mitochondrial large subunit 16S rRNA. J Biol Chem 2014;289:24936–42.

[17] Gong X, Zhang C, Yiliyasi-Aisa Y, Shi Y, Yang XW, Nuersimanguli Aosiman N, Guan YQ, Xu SH. A comparative analysis of genetic diversity of candidate genes associated with type 2 diabetes in worldwide populations. Yi Chuan 2016;38(6):543–59.

[18] Vilardo E, Nachbagauer C, Buzet A, Taschner A, Holzmann J, Rossmanith W. A subcomplex of human mitochondrial RNase P is a bifunctional methyltransferase–extensive moonlighting in mitochondrial tRNA niogenesis. Nucleic Acids Res 2012;40(22):11583–93.

[19] Gillis D, Krishnamohan A, Yaacov B, Shaag A, Jackman EJ, Elpeleg O. TRMT10A dysfunction is associated with abnormalities in glucose homeostasis, short stature and microcephaly. J Med Genet 2014;51(9):581–6.

[20] Igoillo-Esteve M, Genin A, Lambert N, Désir J, Pirson I, Abdulkarim B, Simonis N, Drielsma A, Marselli L, Marchetti P, Vanderhaeghen P, Eizirik DL, Wuyts W, Julier C, Chakera AJ, Ellard S, Hattersley AT, Abramowicz M, Cnop M. tRNA methyltransferase homolog gene TRMT10A mutation in young onset diabetes and primary microcephaly in humans. PLoS Genet 2013;9(10), e1003888.

[21] Zung A, Kori M, Burundukov E, Ben-Yosef T, Tatoor Y, Granot E. Homozygous deletion of TRMT10A as part of a contiguous gene deletion in a syndrome of failure to thrive, delayed puberty, intellectual disability and diabetes mellitus. Am J Med Genet A 2015;167A(12):3167–73.

[22] Brzezniak KL, Bijata M, Szczesny JR, Stepien PP. Involvement of human ELAC2 gene product in 3′ end processing of mitochondrial tRNAs. RNA Biol 2011;8(4):616–26.

[23] Metodiev MD, Thompson K, Alston CL, Morris AAM, He L, Assouline Z, Rio M, Bahi-Buisson N, Pyle A, Griffin H, Siira S, Filipovska A, Munnich A, Chinnery PF, McFarland R, Rötig A, Taylor RW. Recessive mutations in TRMT10C cause defects in mitochondrial RNA processing and multiple respiratory chain deficiencies. Am J Hum Genet 2016;98(5):993–1000.
[24] Goll MG, Kirpekar F, Maggert KA, Yoder JA, Hsieh CL, Zhang X, Golic KG, Jacobsen SE, Bestor TH. Methylation of tRNA asp by the DNA methyltransferase homolog Dnmt2. Science 2006;311:395–8.
[25] Dev RR, Ganji R, Singh SP, Mahalingam S, Banerjee S, Khosla S. Cytosine methylation by DNMT2 facilitates stability and survival of HIV-1 RNA in the host cell during infection. Biochem J 2017;474(12):2009–26.
[26] Cheng JX, Chen L, Li Y, Cloe A, Yue M, Wei J, Watanabe KA, Shammo JM, Anastasi J, Shen QJ, Larson RA, He C, Le Beau MM, Vardiman JW. RNA cytosine methylation and methyltransferases mediate chromatin organization and 5-azacytidine response and resistance in leukaemia. Nat Commun 2018;9(1):1163.
[27] Khan MA, Rafiq MA, Noor A, Hussain S, Flores JV, Rupp V, Vincent AK, Malli R, Ali G, Khan FS, Ishak GE, Doherty D, Weksberg R, Ayub M, Windpassinger C, Ibrahim S, Frye M, Ansar M, Vincent JB. Mutation in NSUN2, which encodes an RNA methyltransferase, causes autosomal-recessive intellectual disability. Am J Hum Genet 2012;90(5):856–63.
[28] Flores JV, Cordero-Espinoza L, Oeztuerk-Winder F, Andersson-Rolf A, Selmi T, Blanco S, Tailor J, Dietmann S, Frye M. Cytosine-5 RNA methylation regulates neural stem cell differentiation and motility. Stem Cell Rep 2017;8(1):112–24.
[29] Ito S, Horikawa S, Suzuki T, Kawauchi H, Tanaka Y, Suzuki T, Suzuki T. Human NAT10 is an ATP-dependent RNA acetyltransferase responsible for N4-acetylcytidine formation in 18 S ribosomal RNA (rRNA). J Biol Chem 2014;289(52):35724–30.
[30] Sharma S, Langhendries JL, Watzinger P, Kötter P, Entian KD, Lafontaine DL. Yeast Kre33 and human NAT10 are conserved 18S rRNA cytosine acetyltransferases that modify tRNAs assisted by the adaptor Tan1/THUMPD1. Nucleic Acids Res 2015;43(4):2242–58.
[31] Arango D, Sturgill D, Alhusaini N, Dillman AA, Sweet TJ, Hanson G, Hosogane M, Sinclair WR, Nanan KK, Mandler MD, Fox SD, Zengeya TT, Andresson T, Meier JL, Coller J, Oberdoerffer S. Acetylation of cytidine in mRNA promotes translation efficiency. Cell 2018;175(7):1872–86.
[32] Larrieu D, Britton S, Demir M, Rodriguez R, Jackson SP. Chemical inhibition of NAT10 corrects defects of laminopathic cells. Science 2014;344(6183):527–32.
[33] Tsai K, Jaguva, Vasudevan AA, Martinez, Campos C, Emery A, Swanstrom R, Cullen BR. Acetylation of cytidine residues boosts HIV-1 gene expression by increasing viral RNA stability. Cell Host Microbe 2020;S1931–3128(20):30293–6.
[34] Bhawe K, Felty Q, Yoo C, Ehtesham NZ, Hasnain SE, Singh VP, Mohapatra I, Roy D. Nuclear respiratory factor 1 (NRF1) transcriptional activity-driven gene signature association with severity of astrocytoma and poor prognosis of glioblastoma. Mol Neurobiol 2020. Online ahead of print.
[35] Tschida BR, Temiz NA, Kuka TP, Lee LA, Riordan JD, Tierrablanca CA, Hullsiek R, Wagner S, Hudson WA, Linden MA, Amin K, Beckmann PJ, Heuer RA, Sarver AL, Yang JD, Roberts LR, Nadeau JH, Dupuy AJ, Keng VW, Largaespada DA. Sleeping beauty insertional mutagenesis in mice identifies drivers of steatosis-associated hepatic tumors. Cancer Res 2017;77(23):6576–88.
[36] Zhang H, Hou W, Wang HL, Liu HJ, Jia XY, Zheng XZ, Zou YX, Li X, Hou L, McNutt MA, Zhang B. GSK-3β-regulated N-acetyltransferase 10 is involved in colorectal cancer invasion. Clin Cancer Res 2014;20(17):4717–29.
[37] Cao Y, Yao M, Wu Y, Ma N, Liu H, Zhang B. N-acetyltransferase 10 promotes micronuclei formation to activate the senescence-associated secretory phenotype machinery in colorectal cancer cells. Transl Oncol 2020; 13(8):100783.
[38] Wu J, Zhu H, Wu J, Chen W, Guan X. Inhibition of N-acetyltransferase 10 using remodelin attenuates doxorubicin resistance by reversing the epithelial-mesenchymal transition in breast cancer. Am J Transl Res 2018; 10(1):256–64.
[39] Liang P, Hu R, Liu Z, Miao M, Jiang H, Li C. NAT10 upregulation indicates a poor prognosis in acute myeloid leukemia. Curr Probl Cancer 2020;44(2):100491.
[40] Zhang X, Chen J, Jiang S, He S, Bai Y, Zhu L, Ma R, Liang X. N-acetyltransferase 10 enhances doxorubicin resistance in human hepatocellular carcinoma cell lines by promoting the epithelial-to-mesenchymal transition. Oxidative Med Cell Longev 2019;2019:7561879.
[41] Songe-Møller L, van den Born E, Leihne V, Vågbø CB, Kristoffersen T, Krokan HE, Kirpekar F, Falnes PØ, Klungland A. Mammalian ALKBH8 possesses tRNA methyltransferase activity required for the biogenesis of multiple wobble uridine modifications implicated in translational decoding. Mol Cell Biol 2010;30 (7):1814–27.

[42] Fu D, Brophy JA, Chan CT, Atmore KA, Begley U, Paules RS, Dedon PC, Begley TJ, Samson LD. Human AlkB homolog ABH8 is a tRNA methyltransferase required for wobble uridine modification and DNA damage survival. Mol Cell Biol 2010;30(10):2449–59.

[43] Ohshio I, Kawakami R, Tsukada Y, Nakajima K, Kitae K, Shimanoe T, Saigo Y, Hase H, Ueda Y, Jingushi K, Tsujikawa K. ALKBH8 promotes bladder cancer growth and progression through regulating the expression of survivin. Biochem Biophys Res Commun 2016;477(3):413–8.

[44] Chen J, Patton JR. Cloning and characterization of a mammalian pseudouridine synthase. RNA 1999;5(3):409–19.

[45] Casas K, Bykhovskaya Y, Mengesha E, Wang D, Yang H, Taylor K, Inbal A, Fischel-Ghodsian N. Gene responsible for mitochondrial myopathy and sideroblastic anemia (MSA) maps to chromosome 12q24.33. Am J Med Genet A 2004;127A(1):44–9.

[46] Bykhovskaya Y, Casas K, Mengesha E, Inbal A, Fischel-Ghodsian N. Missense mutation in pseudouridine synthase 1 (PUS1) causes mitochondrial myopathy and sideroblastic anemia (MLASA). Am J Hum Genet 2004;74(6):1303–8.

[47] Patton JR, Bykhovskaya Y, Mengesha E, Bertolotto C, Fischel-Ghodsian N. Mitochondrial myopathy and sideroblastic anemia (MLASA): missense mutation in the pseudouridine synthase 1 (PUS1) gene is associated with the loss of tRNA pseudouridylation. J Biol Chem 2005;280(20):19823–8.

[48] Fernandez-Vizarra E, Berardinelli A, Valente L, Tiranti V, Zeviani M. Nonsense mutation in pseudouridylate synthase 1 (PUS1) in two brothers affected by myopathy, lactic acidosis and sideroblastic anaemia (MLASA). J Med Genet 2007;44(3):173–80.

[49] Bykhovskaya Y, Mengesha E, Fischel-Ghodsian N. Pleiotropic effects and compensation mechanisms determine tissue specificity in mitochondrial myopathy and sideroblastic anemia (MLASA). Mol Genet Metab 2007; 91(2):148–56.

[50] Fernandez-Vizarra E, Berardinelli A, Valente L, Tiranti V, Zeviani M. Nonsense mutation in pseudouridylate synthase 1 (PUS1) in two brothers affected by myopathy, lactic acidosis and sideroblastic anaemia (MLASA). BMJ Case Rep 2009;2009. bcr05.2009.1889.

[51] Riley LG, Cooper S, Hickey P, Rudinger-Thirion J, McKenzie M, Compton A, Lim SC, Thorburn D, Ryan MT, Giegé R, Bahlo M, Christodoulou J. Mutation of the mitochondrial tyrosyl-tRNA synthetase gene, YARS2, causes myopathy, lactic acidosis, and sideroblastic anemia–MLASA syndrome. Am J Hum Genet 2010;87(1):52–9.

[52] Shahni R, Wedatilake Y, Cleary MA, Lindley KJ, Sibson KR, Rahman S. A distinct mitochondrial myopathy, lactic acidosis and sideroblastic anemia (MLASA) phenotype associates with YARS2 mutations. Am J Med Genet A 2013;161A(9):2334–8.

[53] Burrage LC, Tang S, Wang J, Donti TR, Walkiewicz M, Luchak JM, Chen LC, Schmitt ES, Niu Z, Erana R, Hunter JV, Graham BH, Wong LJ, Scaglia F. Mitochondrial myopathy, lactic acidosis, and sideroblastic anemia (MLASA) plus associated with a novel de novo mutation (m.8969G>A) in the mitochondrial encoded ATP6 gene. Mol Genet Metab 2014;113(3):207–12.

[54] Cao M, Donà M, Valentino ML, Valentino L, Semplicini C, Maresca A, Cassina M, Torraco A, Galletta E, Manfioli V, Sorarù G, Carelli V, Stramare R, Bertini E, Carrozzo R, Salviati L, Pegoraro E. Clinical and molecular study in a long-surviving patient with MLASA syndrome due to novel PUS1 mutations. Neurogenetics 2016;17(1):65–70.

[55] Ozanick S, Krecic A, Andersland J, Anderson TJ. The bipartite structure of the tRNA m1A58 methyltransferase from S. Cerevisiae is conserved in humans. RNA 2005;11(8):1281–90.

[56] Sekar S, McDonald J, Cuyugan L, Aldrich J, Kurdoglu A, Adkins J, Serrano G, Beach TG, Craig DW, Valla J, Reiman EM, Liang WS. Alzheimer's disease is associated with altered expression of genes involved in immune response and mitochondrial processes in astrocytes. Neurobiol Aging 2015;36(2):583–91.

[57] Couch FJ, Kuchenbaecker KB, Michailidou K, Mendoza-Fandino GA, Nord S, Lilyquist J, Olswold C, Hallberg E, Agata S, Ahsan H, Aittomäki K, Ambrosone C, Andrulis IL, Anton-Culver H, Arndt V, Arun BK, Arver B, Barile M, Barkardottir RB, Barrowdale D, Beckmann L, Beckmann MW, Benitez J, Blank SV, Blomqvist C, Bogdanova NV, Bojesen SE, Bolla MK, Bonanni B, Brauch H, Brenner H, Burwinkel B, Buys SS, Caldes T, Caligo MA, Canzian F, Carpenter J, Chang-Claude J, Chanock SJ, Chung WK, Claes KB, Cox A, Cross SS, Cunningham JM, Czene K, Daly MB, Damiola F, Darabi H, de la Hoya M, Devilee P, Diez O, Ding YC, Dolcetti R, Domchek SM, Dorfling CM, Dos-Santos-Silva I, Dumont M, Dunning AM, Eccles DM, Ehrencrona H, Ekici AB, Eliassen H, Ellis S, Fasching PA, Figueroa J, Flesch-Janys D, Försti A, Fostira F, Foulkes WD, Friebel T, Friedman E, Frost D, Gabrielson M, Gammon MD, Ganz PA, Gapstur SM, Garber J, Gaudet MM, Gayther SA, Gerdes AM, Ghoussaini M, Giles GG, Glendon G, Godwin AK, Goldberg MS, Goldgar DE, González-Neira A, Greene MH, Gronwald J, Guénel P, Gunter M, Haeberle L, Haiman CA, Hamann U, Hansen TV, Hart S, Healey S, Heikkinen T, Henderson BE, Herzog J, Hogervorst FB, Hollestelle A, Hooning MJ, Hoover RN, Hopper JL,

Humphreys K, Hunter DJ, Huzarski T, Imyanitov EN, Isaacs C, Jakubowska A, James P, Janavicius R, Jensen UB, John EM, Jones M, Kabisch M, Kar S, Karlan BY, Khan S, Khaw KT, Kibriya MG, Knight JA, Ko YD, Konstantopoulou I, Kosma VM, Kristensen V, Kwong A, Laitman Y, Lambrechts D, Lazaro C, Lee E, Le, Marchand L, Lester J, Lindblom A, Lindor N, Lindstrom S, Liu J, Long J, Lubinski J, Mai PL, Makalic E, Malone KE, Mannermaa A, Manoukian S, Margolin S, Marme F, Martens JW, McGuffog L, Meindl A, Miller A, Milne RL, Miron P, Montagna M, Mazoyer S, Mulligan AM, Muranen TA, Nathanson KL, Neuhausen SL, Nevanlinna H, Nordestgaard BG, Nussbaum RL, Offit K, Olah E, Olopade OI, Olson JE, Osorio A, Park SK, Peeters PH, Peissel B, Peterlongo P, Peto J, Phelan CM, Pilarski R, Poppe B, Pylkäs K, Radice P, Rahman N, Rantala J, Rappaport C, Rennert G, Richardson A, Robson M, Romieu I, Rudolph A, Rutgers EJ, Sanchez MJ, Santella RM, Sawyer EJ, Schmidt DF, Schmidt MK, Schmutzler RK, Schumacher F, Scott R, Senter L, Sharma P, Simard J, Singer CF, Sinilnikova OM, Soucy P, Southey M, Steinemann D, Stenmark-Askmalm M, Stoppa-Lyonnet D, Swerdlow A, Szabo CI, Tamimi R, Tapper W, Teixeira MR, Teo SH, Terry MB, Thomassen M, Thompson D, Tihomirova L, Toland AE, Tollenaar RA, Tomlinson I, Truong T, Tsimiklis H, Teulé A, Tumino R, Tung N, Turnbull C, Ursin G, van Deurzen CH, van Rensburg EJ, Varon-Mateeva R, Wang Z, Wang-Gohrke S, Weiderpass E, Weitzel JN, Whittemore A, Wildiers H, Winqvist R, Yang XR, Yannoukakos D, Yao S, Zamora MP, Zheng W, Hall P, Kraft P, Vachon C, Slager S, Chenevix-Trench G, Pharoah PD, Monteiro AA, García-Closas M, Easton DF, Antoniou AC. Identification of four novel susceptibility loci for oestrogen receptor negative breast cancer. Nat Commun 2016;7:11375.

[58] Śledź P, Jinek M. Structural insights into the molecular mechanism of the m(6)a writer complex. elife 2016;5: e18434.

[59] Wang P, Doxtader AK, Nam Y. Structural basis for cooperative function of Mettl3 and Mettl14 methyltransferases. Mol Cell 2016;63(2):306–17.

[60] Wang X, Feng J, Xue Y, Guan Z, Zhang D, Liu Z, Gong Z, Wang Q, Huang J, Tang C, Zou T, Yin P. Structural basis of N(6)-adenosine methylation by the METTL3-METTL14 complex. Nature 2016;534(7608):575–8.

[61] Beemon K, Keith J. Localization of N6-methyladenosine in the Rous sarcoma virus genome. J Mol Biol 1977; 113(1):165–79.

[62] Gokhale NS, McIntyre ABR, McFadden MJ, Roder AE, Kennedy EM, Gandara JA, Hopcraft SE, Quicke KM, Vazquez C, Willer J, Ilkayeva OR, Law BA, Holley CL, Garcia-Blanco MA, Evans MJ, Suthar MS, Bradrick SS, Mason CE, Horner SM. N6-Methyladenosine in Flaviviridae viral RNA genomes regulates infection. Cell Host Microbe 2016;20(5):654–65.

[63] Lichinchi G, Zhao BS, Wu Y, Lu Z, Qin Y, He C, Rana TM. Dynamics of human and viral RNA methylation during Zika virus infection. Cell Host Microbe 2016;20(5):666–73.

[64] Zhou H, Wang F, Wang H, Chen C, Zhang T, Han X, Wang D, Chen C, Wu C, Xie W, Wang Z, Zhang L, Wang L, Yang H. The conformational changes of Zika virus methyltransferase upon converting SAM to SAH. Oncotarget 2017;8(9):14830–4.

[65] Ye F, Chen ER, Nilsen TW. Kaposi's sarcoma-associated herpesvirus utilizes and manipulates RNA N6-adenosine methylation to promote lytic replication. J Virol 2017;91(16), e00466-17.

[66] Hesser CR, Karijolich J, Dominissini D, He C, Glaunsinger BA. N6-methyladenosine modification and the YTHDF2 reader protein play cell type specific roles in lytic viral gene expression during Kaposi's sarcoma-associated herpesvirus infection. PLoS Pathog 2018;14(4), e1006995.

[67] Lichinchi G, Gao S, Saletore Y, Gonzalez GM, Bansal V, Wang Y, Mason CE, Rana TM. Dynamics of the human and viral m(6)A RNA methylomes during HIV-1 infection of T cells. Nat Microbiol 2016;1:16011.

[68] Courtney DG, Kennedy EM, Dumm RE, Bogerd HP, Tsai K, Heaton NS, Cullen BR. Epitranscriptomic enhancement of influenza a virus gene expression and replication. Cell Host Microbe 2017;22(3):377–86.

[69] Li Z, Shi J, Yu L, Zhao X, Ran L, Hu D, Song B. N6-methyl-adenosine level in Nicotiana tabacum is associated with tobacco mosaic virus. Virol J 2018;15(1):87.

[70] Han M, Liu Z, Xu Y, Liu X, Wang D, Li F, Wang Y, Bi J. Abnormality of m6A mRNA methylation is involved in Alzheimer's disease. Front Neurosci 2020;14:98.

[71] Qin L, Min S, Shu L, Pan H, Zhong J, Guo J, Sun Q, Yan X, Chen C, Tang B, Xu Q. Genetic analysis of N6-methyladenosine modification genes in Parkinson's disease. Neurobiol Aging 2020;93:143. e9-e13.

[72] Liu J, Sun G, Pan S, Qin M, Ouyang R, Li Z, Huang J. The cancer genome atlas (TCGA) based m6A methylation-related genes predict prognosis in hepatocellular carcinoma. Bioengineered 2020;11(1):759–68.

[73] Ghazi T, Nagiah S, Chuturgoon AA. Fusaric acid decreases p53 expression by altering promoter methylation and m6A RNA methylation in human hepatocellular carcinoma (HepG2) cells. Epigenetics 2020;7:1–13.

[74] Huang H, Bai Y, Lu X, Xu Y, Zhao H, Sang X. N6-methyladenosine associated prognostic model in hepatocellular carcinoma. Ann Transl Med 2020;8(10):633.

[75] Lu S, Yu Z, Xiao Z, Zhang Y. Gene signatures and prognostic values of m6A genes in nasopharyngeal carcinoma. Front Oncol 2020;10:875.

[76] Liu ZF, Yang J, Wei SP, Luo XG, Jiang QS, Chen T, Gong YQ. Upregulated METTL3 in nasopharyngeal carcinoma enhances the motility of cancer cells. Kaohsiung J Med Sci 2020; [Online ahead of print].

[77] Xie JW, Huang XB, Chen QY, Ma YB, Zhao YJ, Liu LC, Wang JB, Lin JX, Lu J, Cao LL, Lin M, Tu RH, Zheng CH, Huang CM, Li P. m6A modification-mediated BATF2 acts as a tumor suppressor in gastric cancer through inhibition of ERK signaling. Mol Cancer 2020;19(1):114.

[78] Wang Q, Geng W, Guo H, Wang Z, Xu K, Chen C, Wang S. Emerging role of RNA methyltransferase METTL3 in gastrointestinal cancer. J Hematol Oncol 2020;13(1):57.

[79] Ling Z, Chen L, Zhao J. m6A-dependent up-regulation of DRG1 by METTL3 and ELAVL1 promotes growth, migration, and colony formation in osteosarcoma. Biosci Rep 2020;40(4), BSR20200282.

[80] Zhang QJ, Luan JC, Song LB, Cong R, Ji CJ, Zhou X, Xia JD, Song NH. m6A RNA methylation regulators correlate with malignant progression and have potential predictive values in clear cell renal cell carcinoma. Exp Cell Res 2020;392(1):112015.

[81] Galardi S, Michienzi A, Ciafrè SA. Insights into the regulatory role of m6A Epitranscriptome in glioblastoma. Int J Mol Sci 2020;21(8):2816.

[82] Chen WW, Qi JW, Hang Y, Wu JX, Zhou XX, Chen JZ, Wang J, Wang HH. Simvastatin is beneficial to lung cancer progression by inducing METTL3-induced m6A modification on EZH2 mRNA. Eur Rev Med Pharmacol Sci 2020;24(8):4263–70.

[83] Liang S, Guan H, Lin X, Li N, Geng F, Li J. METTL3 serves an oncogenic role in human ovarian cancer cells partially via the AKT signaling pathway. Oncol Lett 2020;19(4):3197–204.

[84] Zhao W, Cui Y, Liu L, Ma X, Qi X, Wang Y, Liu Z, Ma S, Liu J, Wu J. METTL3 facilitates oral squamous cell carcinoma tumorigenesis by enhancing c-Myc stability via YTHDF1-mediated m6A modification. Mol Ther Nucleic Acids 2020;20:1–12.

[85] Liu L, Wu Y, Li Q, Liang J, He Q, Zhao L, Chen J, Cheng M, Huang Z, Ren H, Chen J, Peng L, Gao F, Chen D, Wang A. METTL3 promotes tumorigenesis and metastasis through BMI1 m(6)a methylation in oral squamous cell carcinoma. Mol Ther 2020;S1525-0016(20):2177–90.

[86] Liu XS, Yuan LL, Gao Y, Zhou LM, Yang JW, Pei ZJ. Overexpression of METTL3 associated with the metabolic status on (18)F-FDG PET/CT in patients with esophageal carcinoma. J Cancer 2020;11(16):4851–60.

[87] Xiang S, Liang X, Yin S, Liu J, Xiang Z. N6-methyladenosine methyltransferase METTL3 promotes colorectal cancer cell proliferation through enhancing MYC expression. Am J Transl Res 2020;12(5):1789–806.

[88] Shen C, Xuan B, Yan T, Ma Y, Xu P, Tian X, Zhang X, Cao Y, Ma D, Zhu X, Zhang Y, Fang JY, Chen H, Hong J. m6A-dependent glycolysis enhances colorectal cancer progression. Mol Cancer 2020;19(1):72.

[89] Zhang J, Bai R, Li M, Ye H, Wu C, Wang C, Li S, Tan L, Mai D, Li G, Pan L, Zheng Y, Su J, Ye Y, Fu Z, Zheng S, Zuo Z, Liu Z, Zhao Q, Che X, Xie D, Jia W, Zeng MS, Tan W, Chen R, Xu RH, Zheng J, Lin D. Excessive miR-25-3p maturation via N6-methyladenosine stimulated by cigarette smoke promotes pancreatic cancer progression. Nat Commun 2019;10(1):1858.

[90] Xia T, Wu X, Cao M, Zhang P, Shi G, Zhang J, Lu Z, Wu P, Cai B, Miao Y, Jiang K. The RNA m6A methyltransferase METTL3 promotes pancreatic cancer cell proliferation and invasion. Pathol Res Pract 2019;215(11):152666.

[91] Konno M, Koseki J, Asai A, Yamagata A, Shimamura T, Motooka D, Okuzaki D, Kawamoto K, Mizushima T, Eguchi H, Takiguchi S, Satoh T, Mimori K, Ochiya T, Doki Y, Ofusa K, Mori M, Ishii H. Distinct methylation levels of mature microRNAs in gastrointestinal cancers. Nat Commun 2019;10(1):3888.

[92] Jia G, Fu Y, Zhao X, Dai Q, Zheng G, Yang Y, Yi C, Lindahl T, Pan T, Yang YG, He C. N6-methyladenosine in nuclear RNA is a major substrate of the obesity-associated FTO. Nat Chem Biol 2011;7(12):885–7.

[93] Jia G, Yang CG, Yang S, Jian X, Yi C, Zhou Z, He C. Oxidative demethylation of 3-methylthymine and 3-methyluracil in single-stranded DNA and RNA by mouse and human FTO. FEBS Lett 2008;582(23–24):3313–9.

[94] Zhou D, Liu H, Zhou M, Wang S, Zhang J, Liao L, He F. Common variant (rs9939609) in the FTO gene is associated with metabolic syndrome. Mol Biol Rep 2012;39(6):6555–61.

[95] Hotta K, Kitamoto T, Kitamoto A, Mizusawa S, Matsuo T, Nakata Y, Kamohara S, Miyatake N, Kotani K, Komatsu R, Itoh N, Mineo I, Wada J, Yoneda M, Nakajima A, Funahashi T, Miyazaki S, Tokunaga K, Masuzaki H, Ueno T, Hamaguchi K, Tanaka K, Yamada K, Hanafusa T, Oikawa S, Yoshimatsu H, Sakata T, Matsuzawa Y,

Nakao K, Sekine A. Association of variations in the FTO, SCG3 and MTMR9 genes with metabolic syndrome in a Japanese population. J Hum Genet 2011;56(9):647–51.

[96] Prakash J, Srivastava N, Awasthi S, Agarwal CG, Natu SM, Rajpal N, Mittal B. Association of FTO rs17817449 SNP with obesity and associated physiological parameters in a north Indian population. Ann Hum Biol 2011;38(6):760–3.

[97] Grunnet LG, Brøns C, Jacobsen S, Nilsson E, Astrup A, Hansen T, Pedersen O, Poulsen P, Quistorff B, Vaag A. Increased recovery rates of phosphocreatine and inorganic phosphate after isometric contraction in oxidative muscle fibers and elevated hepatic insulin resistance in homozygous carriers of the A-allele of FTO rs9939609. J Clin Endocrinol Metab 2009;94(2):596–602.

[98] Sabarneh A, Ereqat S, Cauchi S, AbuShamma O, Abdelhafez M, Ibrahim M, Nasereddin A. Common FTO rs9939609 variant and risk of type 2 diabetes in Palestine. BMC Med Genet 2018;19(1):156.

[99] Li Y, Zheng D, Wang F, Xu Y, Yu H, Zhang H. Expression of demethylase genes, FTO and ALKBH1, is associated with prognosis of gastric cancer. Dig Dis Sci 2019;64(6):1503–13.

[100] Zou D, Dong L, Li C, Yin Z, Rao S, Zhou Q. The m(6)a eraser FTO facilitates proliferation and migration of human cervical cancer cells. Cancer Cell Int 2019;19:321.

[101] Liu Y, Wang R, Zhang L, Li J, Lou K, Shi B. The lipid metabolism gene FTO influences breast cancer cell energy metabolism via the PI3K/AKT signaling pathway. Oncol Lett 2017;13(6):4685–90.

[102] Niu Y, Lin Z, Wan A, Chen H, Liang H, Sun L, Wang Y, Li X, Xiong XF, Wei B, Wu X, Wan G. RNA N6-methyladenosine demethylase FTO promotes breast tumor progression through inhibiting BNIP3. Mol Cancer 2019;18(1):46.

[103] Yang S, Wei J, Cui YH, Park G, Shah P, Deng Y, Aplin AE, Lu Z, Hwang S, He C, He YY. m(6)A mRNA demethylase FTO regulates melanoma tumorigenicity and response to anti-PD-1 blockade. Nat Commun 2019;10(1):2782.

[104] Huang Y, Su R, Sheng Y, Dong L, Dong Z, Xu H, Ni T, Zhang ZS, Zhang T, Li C, Han L, Zhu Z, Lian F, Wei J, Deng Q, Wang Y, Wunderlich M, Gao Z, Pan G, Zhong D, Zhou H, Zhang N, Gan J, Jiang H, Mulloy JC, Qian Z, Chen J, Yang CG. Small-molecule targeting of oncogenic FTO demethylase in acute myeloid leukemia. Cancer Cell 2019;35(4):677–91.

[105] Zheng G, Dahl JA, Niu Y, Fedorcsak P, Huang CM, Li CJ, Vågbø CB, Shi Y, Wang WL, Song SH, Lu Z, Bosmans RP, Dai Q, Hao YJ, Yang X, Zhao WM, Tong WM, Wang XJ, Bogdan F, Furu K, Fu Y, Jia G, Zhao X, Liu J, Krokan HE, Klungland A, Yang YG, He C. ALKBH5 is a mammalian RNA demethylase that impacts RNA metabolism and mouse fertility. Mol Cell 2013;49(1):18–29.

[106] Tang B, Yang Y, Kang M, Wang Y, Wang Y, Bi Y, He S, Shimamoto F. m(6)A demethylase ALKBH5 inhibits pancreatic cancer tumorigenesis by decreasing WIF-1 RNA methylation and mediating Wnt signaling. Mol Cancer 2020;19(1):3.

[107] Cho SH, Ha M, Cho YH, Ryu JH, Yang K, Lee KH, Han ME, Oh SO, Kim YH. ALKBH5 gene is a novel biomarker that predicts the prognosis of pancreatic cancer: a retrospective multicohort study. Ann Hepatobiliary Pancreat Surg 2018;22(4):305–9.

[108] Strick A, von Hagen F, Gundert L, Klümper N, Tolkach Y, Schmidt D, Kristiansen G, Toma M, Ritter M, Ellinger J. The N(6) -methyladenosine (m(6) a) erasers alkylation repair homologue 5 (ALKBH5) and fat mass and obesity-associated protein (FTO) are prognostic biomarkers in patients with clear cell renal carcinoma. BJU Int 2020;125(4):617–24.

[109] Zhang C, Zhi WI, Lu H, Samanta D, Chen I, Gabrielson E, Semenza GL. Hypoxia-inducible factors regulate pluripotency factor expression by ZNF217- and ALKBH5-mediated modulation of RNA methylation in breast cancer cells. Oncotarget 2016;7(40):64527–42.

[110] Zhu H, Gan X, Jiang X, Diao S, Wu H, Hu J. ALKBH5 inhibited autophagy of epithelial ovarian cancer through miR-7 and BCL-2. J Exp Clin Cancer Res 2019;38(1):163.

[111] Zhang J, Guo S, Piao HY, Wang Y, Wu Y, Meng XY, Yang D, Zheng ZC, Zhao Y. ALKBH5 promotes invasion and metastasis of gastric cancer by decreasing methylation of the lncRNA NEAT1. J Physiol Biochem 2019; 75(3):379–89.

CHAPTER

7

The combination strategy with epigenetic therapy, chemotherapy, and immunotherapy

Shijun Wen[a], *Haiwen Wang*[a], *and Sixiao Lu*[a,b]

[a]State Key Laboratory of Oncology in South China, Sun Yat-sen University Cancer Center, Guangzhou, People's Republic of China [b]Division of Biosciences, University College London, London, United Kingdom

Abstract

Chemotherapy, including targeted chemotherapy, is most commonly used in cancer treatment while immunotherapy has appeared therapeutically promising recently. The common issues concerning these two therapeutic approaches include low response and drug resistance. Moreover, monotherapy is not sufficient to eliminate cancer cells completely due to multiple factors contributing to tumor malignancy. Abnormal epigenetic reprogramming mediates drug interference and promotes cancer cell survival and resistance. Epigenetic drugs intervening in such reprograming, including inhibitors of histone deacetylase (HDAC), bromodomain and extra-terminal (BET) family proteins, histone methyltransferase EZH2, lysine-specific demethylase (LSD1), and DNM methyltransferase (DNMT) may overcome drug resistance. Therefore, the combination of epigenetic drugs with either chemotherapy or immunotherapy may have advantages over monotherapy. In this chapter, we elucidate the role epigenetic progress plays in drug resistance and review the advancements in epigenetic therapy combined with chemotherapy and immunotherapy to enhance drug response and improve therapeutic efficacy.

Abbreviations

AML	acute myeloid leukemia
APL	acute promyelocytic leukemia
AR	androgen receptor
ATRA	all-trans-retinoic acid
BET	bromodomain and extra-terminal
BRCA	breast cancer gene
BRD4	bromodomain-containing protein 4
CDK	cyclin-dependent kinase

https://doi.org/10.1016/B978-0-12-823684-0.00013-X

CRPC	castration-resistant prostate cancer
CTLA-4	cytotoxic T-lymphocyte-associated protein 4
CXCL8	interleukin-8
DHL	double hit lymphoma
DNMT	DNM methyl transferase
DSB	DNA double-strand break
dsRNA	double-stranded RNA
EGFR	epidermal growth factor receptor
EIR	EZH2 inhibitor-resistant
EMT	epithelial-mesenchymal transition
ERα	estrogen receptor α
ERVs	endogenous retroviral elements
EZH2	enhancer of zeste homolog 2
HCC	hepatocellular carcinoma
HDAC	histone deacetylase
HNC	head and neck cancer
HR	homologous recombination
IL6R	IL6 receptor
LSD1	lysine-specific demethylase 1
MCL	mantle cell lymphoma
MDS	myelodysplastic syndrome
MDSC	myeloid derived suppressor cell
mTORC1	mammalian target of rapamycin complex 1
NSCLC	nonsmall cell lung cancer
PARP	poly (ADP-ribose) polymerase
PCa	prostate cancer
PEL	primary effusion lymphoma
PD1	programmed cell death-1
PD-L1	programmed death-ligand 1
RAR	retinoic acid receptor
RB	retinoblastoma
RISC	RNA-induced silencing complex
sAML	secondary acute myeloid leukemia
TCP	tranylcypromine
TKIs	tyrosine kinase inhibitors
TMB	tumor mutation burden
TME	tumor microenvironment
TNBC	triple-negative breast cancer
TNF-α	tumor necrosis factor α

Conflict of interest

No potential conflicts of interest were disclosed.

Introduction

In the fight against cancer, traditional chemotherapy with unselective cytotoxic drugs including DNA damaging agents and antimicrotubule agents has undoubtedly exerted its substantial effect to kill cancer cells and suppress cancer development. However, traditional chemotherapy inevitably faces challenges of disease relapse, drug resistance, and other side effects. Therefore, advanced alternative therapeutic approaches to eliminate cancer cells are in demand [1]. The increasing knowledge of molecular biology and oncological biology has given better understanding in tumorigenesis, offering a scientific basis to seek more effective and novel therapeutic strategies. Targeted small molecular drugs selectively sniping at particular proteins, which are dysregulated and determinant in tumorigenesis and cancer proliferation, emerge as one of the appealing alternative options to achieve cancer elimination [2]. Various chemotherapy drugs targeting individual pathways have been well developed and show their satisfactory effect on suppressing cancer; however, they fail to avoid drug resistance which seriously limits their clinical usage [3]. The reasons for primary or acquired drug resistance can be complex, including cancer stem cell existence, drug transportation barrier, drug inactivation, target mutation, and altered target expression at DNA, mRNA, or protein levels [4, 5]. Identifying the key factors contributing to the response to chemotherapy drugs and regulating these factors by adding another targeted drug may provide a solution to overcome drug resistance and enhance drug efficacy, which leads to the development of a drug combination strategy [6].

Immunotherapy is burgeoning as a very promising therapeutic approach to cure cancer [7]. The tumor microenvironment has been under intensive investigation to elucidate how cancer cells impair immune response and thus escape from immune surveillance, including suppressing chemokine expression and infiltration, inactivating T cells, inducing immune cell differentiation and exhaustion [8]. Cancer immunosuppression can result from genetic or epigenetic alterations [9, 10]. Liberating immunosuppression and harnessing cytotoxic immune cells to kill cancer cells become the basic rule of immunotherapy and the foundation in drug development [11]. Taking PD-1 and PD-L1 for instance, cancer cells express PD-L1 that can interact with PD-1 on T cells to cause inactivation of T cells [12]. Such interaction can be intercepted by anti-PD-1/PD-L1 agents [13]. As a result, the inactivation of T cells is removed and the functional T cells restore their anticancer effect. PD-1/PD-L1 are recognized as promising targets for immunotherapy, and anti-PD-1/PD-L1 drugs have been developed [14]. Although immunotherapy has significant improvement in curing cancer, the efficacy of these immune-checkpoint blockages varies in different types of cancers due to multiple immunosuppressive mechanisms under a special tumor microenvironment [15]. The acquired resistance during continuous heightened immune pressure makes things worse [16]. In many cases, anti-PD-1 inhibitor alone is insufficient to suppress tumor growth, suggesting that additional medication is needed to improve immune response and prevent resistance development [17].

Abnormal epigenetic processes that regulate the states of histone acetylation, methylation, DNA methylation, and so on are often observed in cancer cells. Epigenetic therapy and epigenetic inhibitors gain attention not only due to their remarkable curative effect on suppressing tumors alone [18, 19], but also due to their potency in combination with other therapies [20]. Epigenetic reprogramming is recognized as one of the major ways to mediate drug interference and promote cancer cell survival and resistance [21, 22]. Through altered transcriptional levels, cancer cells downregulate targeted proteins or upregulate pro-survival proteins to eliminate drug influence. Epigenetic inhibitors intervening in such abnormal reprograming may overcome drug resistance and enhance drug efficacy. Epigenetic progress also plays a significant role in cancer immune responses [23]. The efficiency of cancer immunotherapy is often decided by several factors including tumor mutation burden (TMB), state of immune cells, tumor microenvironment (TME), and mechanisms of tumor immune evasion [24]. Epigenetic progresses in both cancer cells and immune cells play an important role in impairing immune response, resisting immune-checkpoint blockage and promoting cell survival. Epigenetic inhibitors reversing these negative progresses can improve immune response and augment the effect of immune-checkpoint blockage treatment. Currently, the inhibitors of histone deacetylase (HDAC), bromodomain and extra-terminal (BET) family proteins, histone methyltransferase EZH2, lysine-specific demethylase (LSD1), and DNM methyltransferase (DNMT) are either approved in clinical therapy or under clinical trials.

Taken together, there have been dramatic advancements in cancer treatment using chemotherapy, immunotherapy, or epigenetic therapy alone. However, due to the intrinsic defect of each therapy, a combination therapy may have therapeutic advantages. Indeed, a combination of chemotherapy, immunotherapy, and epigenetic therapy elevates drug sensitivity, retards acquired resistance, and suppresses tumors more effectively, providing a promising approach for cancer treatment.

Epigenetic therapy combining with chemotherapy

Epigenetic progress plays a significant role in regulating expression of genes that are important to cell survival and growth. Aberrant epigenetic progress not only induces cancer malignance, but also impairs tumor sensitivity to chemotherapy. Through epigenetic alteration, cancer cells regulate relevant signaling pathways to offset and/or resist the effects caused by chemotherapy, gaining acquired drug resistance gradually. Thus, epigenetic drugs that can suppress these epigenetic alterations are helpful to enhance the efficacy of chemotherapy and overcome the acquired drug resistance, providing a foundation that a combination of epigenetic inhibitors and chemotherapy may elicit a synergistic effect to cure cancer.

Combination of HDAC inhibitors and chemotherapy

Acquired resistance against platinum drugs becomes an obstacle to cisplatin-based chemotherapy for advanced nonsmall cell lung cancer (NSCLC). Multiple mechanisms are involved in this acquired drug resistance, including the accumulation of cancer stem cells [25]. Evidences support that cancer stem cells result from the reprogrammed epigenetic progress that silences gene expression [26]. Importantly, it has been found that the continuous treatment with cisplatin can result in HDAC activation, which may contribute to cancer stem cell accumulation and drug resistance. HDAC inhibition can prevent the development of cisplatin resistance and offers an ideal candidate for combination to enhance cisplatin-based chemotherapy. In practice, the combination of HDAC inhibitors and cisplatin synergistically inhibits tumor growth in a cisplatin-resistant NCI-H460 xenograft model [27].

In cancer cells, the aberration of DNA repair genes is associated with the sensitivity to DNA damage drugs such as poly(ADP-ribose) polymerase (PARP) inhibitors [28]. BRCA1 and BRCA2, two DNA repair genes, are critical tumor suppressor genes for DNA double-strand break (DSB) repair through homologous recombination (HR) pathways [29]. BRCA can attenuate the DNA damage effect caused by PARP inhibitors and make cancer cells less sensitive to PARP inhibition. Interestingly, HDAC is reportedly involved in HR DNA repair [30]. HDAC inhibitors successfully prevent the expression of BRCA1 and subsequently enhance sensitivity to PARP inhibitors. Combinations of HDAC inhibitors and PARP inhibitors have gained considerable interest. HDAC inhibitor SAHA and PARP inhibitor veliparib showed synergistic effects on DNA damage and death of prostate cancer cells by decreasing the protein levels of BRCA1, without affecting normal prostate epithelial cells. Coadministration of both drugs synergistically decreased the growth of tumor xenografts derived from DU145 cells [31].

Epidermal growth factor receptor (EGFR) inhibitors have been used for treating head and neck cancer (HNC) and lung cancer; however, the drug efficacy is limited by primary and acquired resistance which partially results from increased epithelial to mesenchymal transition (EMT) [32]. Interestingly, it is uncovered that HDAC inhibition can reverse EMT in different tumors including HNC [33]. BIM (BCL2L11) with a BH3 domain, a proapoptotic member of the Bcl-2 protein family, is regarded as an essential effector for apoptotic progress, and reduced BIM expression is positively related to EGFR inhibitor resistance in EGFR-mutated nonsmall cell lung cancer (NSCLC) [34]. However, HDAC inhibition is capable of mediating histone hyperacetylation of the promoter region and consequently activating transcription of proapoptotic BH3-only genes including BIM [35]. Thus, the addition of HDAC inhibitors to EGFR inhibitor treatments is hypothesized. Simona Citro et al. report that a combination of SAHA and an EGFR inhibitor gefitinib has a synergistic antitumor activity effect on HNC cell lines [36]. Seiji Yano et al. demonstrate the combination effect of the HDAC inhibitor vorinostat and EGFR inhibitor osimertinib in inducing apoptosis in EGFR-T790M mutation-positive NSCLC cells with BIM deletion polymorphism [37]. Xiufeng Pang et al. report that the combination of the HDAC inhibitor YF454A and EGFR inhibitor erlotinib synergistically inhibits the growth of primary EGFR inhibitor-resistant A549 xenografts [38]. All these preclinical data support that cotreatment with the HDAC inhibitor and EGFR inhibitor is a promising strategy to cure cancers.

As mentioned above, HDAC inhibitors can overcome drug resistance induced by chemotherapy including targeted treatment. On the other hand, clinical studies found that solid cancers show less vulnerability to HDAC inhibitors compared to hematological malignancies, which is disappointing in the epigenetic therapy field [39]. The resistance of solid cancers to HDAC inhibition requires further elucidation. Several works uncover that HDAC inhibition promotes IκB kinase (IKK)-dependent expression of proinflammatory chemokines such as interleukin-8 (CXCL8), resulting in the increased proliferation of tumor cells and accounting for the limited effect in solid cancers [40]. Preventing the HDAC inhibition-induced IKK-dependent CXCL8 expression by IKK inhibitors may help to improve the therapeutic effect of the HDAC inhibitor in ovarian cancer and other solid tumors characterized by increased CXCL8 expression. The combined therapeutic effect with the HDAC inhibitor and IKK inhibitor has been also evaluated in hematological malignancy models. Combining HDAC inhibitor romidepsin and IKK inhibitor LY2409881 produces a remarkable synergistic effect in suppressing diffuse large B cell lymphoma cell growth in vitro and in vivo [41].

Combination of BET inhibitors and chemotherapy

BET family proteins including BRD4 recognize the acetylated lysine tails on histone of which the acetylation state is controlled by HDACs and histone acetyl transferases.

Numerous works have reported that BET proteins play a crucial role in regulating the JAK-STAT signaling pathway. BET inhibitors can interfere with epigenetic reprogramming in the JAK-STAT pathway and thus enhance drug sensitivity and overcome drug resistance to JAK inhibitors [42]. IL6 receptor (IL6R), a direct target of BRD4, presents a strongly high level of expression in several multiple myeloma subgroups [43]. It is found that BET inhibitors like INCB054329 are able to suppress the IL6-JAK-STAT axis through reducing the IL6R transcription level. Cocombating the JAK-STAT pathway with BET inhibitors and JAK inhibitors may provide a promising prospect to cure cancer. Matthew C. Stubbs et al. report that a combination of a JAK inhibitor ruxolitinib or itacitinib and a BET inhibitor INCB054329 exhibits a synergistic effect to inhibit myeloma cell growth in vitro and in vivo. In the MM1.S model of myeloma, which is not intrinsically sensitive to JAK inhibition alone, this combination therapy achieves a remarkable tumor growth inhibition [44]. KN Bhalla et al. report that cotreatment with ruxolitinib, or SAR302503 and another BET inhibitor JQ1, synergistically suppresses the growth of cultured and patient-derived secondary acute myeloid leukemia (sAML) cells and significantly improves the survival of immune-depleted mice engrafted with human sAML cells. Such a combination greatly inhibits prosurvival oncoproteins p-STAT5, p-STAT3, c-MYC, BCL-2, and CDK4/6 [45]. In addition to regulating JAK-STAT axial genes and proteins directly, BET inhibitors regulate other lethal pathways, cooperating with JAK inhibition to enhance an anticancer effect. It is reported that the collaboration between JAK/STAT and NF-kB pathways contributes to aberrant cytokine production and myelofibrosis progression in myeloproliferative neoplasms [46]. BET proteins show a potential function in regulating NF-kB activation, and BET inhibitors are able to attenuate NF-kB transactivation. Combining ruxolitinib and JQ1 successfully reduces inflammation and the disease burden, completely reversing reticulin fibrosis in MF mice [47].

As a BET family protein, BRD4 is essential for the transcriptional activity of NF-kB, and the inhibitor JQ1 is able to inhibit the expression of NF-kB targeted genes including Bruton tyrosine kinase (BTK). Cotreatment with JQ1 and the BTK inhibitor ibrutinib exerts synergistic lethal activity against mantle cell lymphoma (MCL) [48]. Cotreatment with JQ1 and ibrutinib synergizes strongly in killing an activated B-cell-like subtype of diffuse large B-cell lymphoma cells in vitro and in a xenograft mouse model [49].

Meanwhile, some works indicate that the JAK-STAT signaling pathway may be responsible for the sensitivity to cisplatin in tumor cells [50]. Thus, combining BET inhibitors that can suppress the JAK-STAT axial signal with cisplatin becomes a potential therapeutic strategy. Aristotle Bamias et al. use this combination therapy with JQ1 and cisplatin to achieve a synergetic effect in platinum-resistant ovarian cells [51].

Venetoclax, which targets BCL-2 and is approved to clinically treat lymphoid neoplasm, is found ineffective in suppressing $MYC^{+}/BCL\text{-}2^{+}$ double hit lymphoma (DHL) tumor growth. The drug resistance to the BCL-2 inhibitor partly results from the accumulation of a BCL-2 family protein BFL-1 that is a target of BET protein [52, 53]. Thus, BET inhibitors are promising to overcome the drug resistance to BCL-2 inhibitors. The combination therapies using the BCL-2 inhibitor and BET inhibitors have been investigated. A combination of venetoclax and a BET inhibitor ABBV-075 exhibits a synergistic effect across acute myeloid leukemia (AML) cell lines including SKM-1, THP-1, AML5, MV4:11, ML-2, MOLM-13, AML-2, Kasumi, and HL-60 [54]. A combination of venetoclax and another BET inhibitor CPI203 induces synergistic caspase-mediated apoptosis in DHL cells and their tumor xenografts [55].

A combination of venetoclax and JQ1 gains a superior therapeutic effect in mice xenografted with luciferase-positive human LOUCY cells and shows a potency to cure T-cell acute lymphoblastic leukemia [45]. All these evidences support the combined use of BET inhibitors and BCL-2 inhibitors as a promising therapeutic strategy to cure cancer.

BET inhibitors themselves also are in the dilemma of acquired drug resistance via multiple mechanisms [56]. Combination strategies are needed to overcome the resistance to BET inhibitors and exert better effects to cure cancer. The genes and proteins whose expression is reprogrammed during the lasting incubation with BET inhibitors are analyzed. Shu et al. report that reprogramming G1-S progression regulators is responsible for the sensitivity to BET inhibitors. Downregulating positive regulators of G1-S progression including CDK4 and CCND1 enhances the drug sensitivity, providing a basis for a combination of BET inhibitors and CDK4/6 inhibitors. It was later proved that JQ1 treatment induces retinoblastoma (RB) phosphorylation, whereas CDK4/6 inhibitor palbociclib can decrease RB phosphorylation, leading to CDK4 inhibition mediated G1 arrest and reinforcing the sensitivity to JQ1. The combination of palbociclib and JQ1 significantly decreases the tumor growth of resistant SUM159 and IDC50 xenografts [57]. The emerging resistance to BET inhibitors is also the result of the dysfunctional BRD4 protein degradation mechanism in cancers. Deubiquitinase DUB3-proficient prostate cancer cells are resistant to JQ1 because DUB3 is able to deubiquitinate and consequently stabilize BRD4. The strategy to combine JQ1 and DUB3 inhibitory agents such as palbociclib overcomes the JQ1 resistance and gains a synergistic suppressive effect in the DU145 xenograft model [58].

Combination of EZH2 inhibitors and chemotherapy

EZH2 plays a role in acquired resistance against platinum drugs in ovarian cancer. It is implied that a potential mechanism for the development of resistance may lie on the activation of the c-Myc-miR-137-EZH2 pathway induced by cisplatin treatment. EZH2 activates cellular survival pathways, resulting in the resistance to cisplatin. Such resistance can be reversed by EZH2 inhibitors that are able to target the c-Myc-miR-137-EZH2 pathway. EZH2 inhibitors sensitize ovarian cancer cells to cisplatin, and the combined treatment with EZH2 inhibitor GSK343 and cisplatin achieves an excellent synergistic suppression in resistant ovarian cancer cells IGROV1 CR and POE4 [59].

Triple-negative breast cancer (TNBC) is deficient in estrogen receptor α (ERα) and less sensitive to hormone therapy currently used in breast cancer. It is suggested that the CDK2/EZH2 axis regulates the ESR1 gene that encodes ERα [60]. The CDK2-mediated site-specific phosphorylation of EZH2 drives tumorigenesis and contributes to the TNBC phenotype. Inhibition of the CDK2/EZH2 axis allows reexpression of ERα, sensitizing TNBC cells to tamoxifen, an ER inhibitor. The combination of EZH2 inhibitors and tamoxifen effectively suppresses tumor growth and markedly improves the survival of mice bearing TNBC tumors [61], suggesting that EZH2 inhibitors and tamoxifen combination therapy may be an alternative approach to treat TNBC.

The similar predicament persecutes patients with prostate cancer (PCa). While small molecular inhibitors targeting the androgen receptor (AR) serve as the major therapy for PCa patients, unfortunately about 30% of patients possess primary resistance to AR inhibitors [62].

Furthermore, the majority of patients progress from androgen-dependent PCa to castration-resistant PCa (CRPC). The opportunity to overcome the resistance to AR inhibitors and enhance the AR-inhibitor-based therapy may begin with the discovery that AR is a direct target of EZH2-mediated transcriptional activation. EZH2 can occupy the AR promoter to directly induce its transcription. The upregulation of the AR level and resistance to AR inhibitors can be distinctly reversed by EZH2 inhibition, which supports the strategy combining the two inhibitors of EZH2 and AR. The combination treatment significantly reduces the tumor growth of CRPC xenografts while each inhibitor alone fails the therapeutic efficacy [63].

PARP inhibitors are promising targeted agents to treat BRCA-mutant ovarian cancer. However, PARP inhibition treatment can promote EZH2-mediated gene silencing and cancer stem cell properties through suppressing alkylating DNA damage-induced PARylation. Using EZH2 inhibitor is successful to offset this countereffect and sensitizes the BRCA-mutant breast cells to PARP inhibitors. Combining the EZH2 inhibitor GSK343 and olaparib significantly inhibits tumor growth in a SUM149 xenograft mouse model [64].

EZH2 inhibitors also face a challenge of the acquired drug resistance, which limits their therapeutic efficacy. The study of their mechanisms helps to find relevant switches to turn down such drug resistance, furthermore directing combination approaches with EZH2 inhibitors. The chromatin remodeling complex SWI/SNF catalytic subunit switching from SMARCA4 to SMARCA2 drives resistance to EZH2 inhibitors in ARID1A-mutated cells [65]. Such a switch leads to suppression of apoptotic pathways through upregulating antiapoptotic genes such as BCL2 in the EZH2 inhibitor-resistant (EIR) cells. As a result, EIR ARID1A-mutated cells are hypersensitive to BCL2 inhibitors such as ABT263. ABT263 shows the ability to overcome resistance to the EZH2 inhibitor and exhibits a synergistic effect with EZH2 inhibitors in suppressing the tumor growth of xenografts of ARID1A-deficient primary OCCC cells [66].

Combination of LSD1 inhibitors and chemotherapy

Lysine-specific histone demethylase 1 (LSD1) plays a critical role in maintaining a pool of tumor-initiating cells that may contribute to the development of chemoresistance. Targeting LSD1 in a combination with conventional therapeutic methods could be more efficacious to treat cancer than monotherapy alone. Verigos et al. demonstrate that the pretreatment with LSD1 inhibitors renders breast cancer cells more sensitive to a common chemotherapeutic drug doxorubicin, leading to a more significant reduction of tumor cells [67]. Wen Peng et al. report that ZY0511, a novel LSD1 inhibitor, can potentiate the anticancer effects of another traditional chemotherapeutic drug 5-fluorouracil and synergistically suppresses the growth and metastasis of colorectal cancer cells [68].

All-trans-retinoic acid (ATRA)-based treatment is effective in treating a subtype of AML called acute promyelocytic leukemia (APL) by promoting the differentiation of abnormal blast cells into normal white blood cells. However, ATRA resistance appears in patients with non-APL AML while ATRA fails to induce transcriptional activation of retinoic acid receptor (RAR) target genes including TNFSF10 and RARA2. In this case, inducing epigenetic reprogramming to increase H3K4 dimethylation on the promoter region of RAR2 could upregulate the expression of genes associated with myeloid differentiation. The treatment of ATRA plus an irreversible LSD1 inhibitor tranylcypromine (TCP) is found to target leukemia-initiating cells and has a more potent antileukemic effect than each single treatment alone [69]. Recently, Smitheman et al. report that combining ATRA and another LSD1 inhibitor GSK2879552 can enhance cell differentiation of AML and, more importantly, induce a synergistic cytotoxicity that leads to AML cell elimination [70]. These studies have demonstrated that LSD1 inhibitors can expand the therapeutic potential of ATRA, paving the way for new approaches in treating AML. The combination of TCP and ATRA for treating relapsed/refractory AML and myelodysplastic syndrome (MDS) has already been tested in the phase I clinical trial, demonstrating an acceptable safety profile and antileukemic clinical activity [71]. Currently, two LSD1 inhibitors under ongoing clinical trials on testing the maximum tolerant dose are IMG-7289 or TCP in a combination of ATRA in patients with AML and MDS (ClinicalTrials.gov identifier: NCT02842827 and NCT02717884). The results from these clinical trials will help to identify the optimal doses of LSD1 inhibitors while treating the patients with combination therapy. Further investigations on clinical trials will explore different schedules or better patient selection.

LSD1 has emerged as a candidate therapeutic target in AML, and LSD1 inhibitors induce loss of clonogenic activity and promote the transdifferentiation of AML cells. However, AML cells frequently exhibit intrinsic or acquired resistance to LSD1 inhibition. In this case, combination regimens can help to overcome the resistance to LSD1 inhibitors. Mammalian target of rapamycin complex 1 (mTORC1) signaling plays a role in the activation and differentiation of myeloid cells and can be triggered in resistant leukemia following LSD1 inhibition. It is believed that mTOR is involved in mediating the resistance of leukemia cells to LSD1 inhibitors. Deb et al. demonstrate that combined pharmacologic inhibition of mTORC1 and LSD1 not only enhances myeloid differentiation but also impairs AML cell growth [72]. Moreover, Abdel-Aziz et al. indicate that inhibiting mTOR sensitizes tolerant AML cell lines and primary patient-derived AML blasts to LSD1 inhibitors [73]. Overall, coinhibiting LSD1 and mTOR restores the efficacy of LSD1 inhibitors in irresponsive AML patients.

LSD1 inhibitors can also synergize with targeted chemotherapies, for example, the use of tyrosine kinase inhibitors (TKIs). Most patients with NSCLC eventually develop resistance to epidermal growth factor receptor (EGFR) TKIs after NSCLCs elicit secondary mutations in EGFR or other downstream signaling components. Histone demethylases are reportedly involved in the development of EGFR TKIs resistance. Lu et al. have reported that LSD1 is required for hypoxia-induced EGFR TKI resistance in NSCLC cells [74]. Thus, LSD1 inactivation could prevent the drug resistance acquisition or resensitize the resistant cancer cells to EGFR TKIs. According to their study, a combination of an LSD1 inhibitor SP2509 and gefitinib significantly inhibits NSCLC tumor growth [65], thus providing a preclinical implication for the use of LSD1 inhibitors in treating NSCLC. Furthermore, the combination of LSD1 inhibitors and a TKI sorafenib can also exert a synergistic anticancer effect on patients with advanced hepatocellular carcinoma (HCC). Acquired drug resistance frequently occurs in HCC patients treated by sorafenib, and it is mainly induced by the enrichment of self-renewing cancer stem cells. The stem-like properties of sorafenib-resistant HCC cells can be suppressed by LSD1 inhibitors that derepress the expression of the Wnt/β-catenin signaling pathway. In this way, LSD1 inhibition resensitizes sorafenib-resistant HCC cells to sorafenib [75]. Regorafenib, a multikinase inhibitor, triggers acquired drug resistance by activating the AKT/FOXM1/STMN1 pathway while LSD1 abrogation alleviates the activated p-AKT in HCC cells. Thus, SP2509 and TCP enhance the cytotoxic and apoptotic effects of regorafenib in HCC cells [76].

Combination of DNMT inhibitors and chemotherapy

The mechanisms underlying the development of chemotherapy resistance can be complex, including the intrinsic stemness properties of cancer cells. Bruno Silva-Santos et al. found that DNMT inhibitors prevent chemoresistant relapses by suppressing the outgrowth of a set of chemoresistant AML clones with stemness properties. The combination of a DNMT inhibitor decitabine and chemotherapy drugs including doxorubicin and cytarabine can prevent chemoresistance in human AML cell lines HEL and OCI-AML3 [77].

DNMT inhibitors are a also used to reverse intrinsic or acquired resistance to the PARP inhibitor. By increasing ROS accumulation and promoting the cAMP/PKA pathway and PARP activation, DNMT inhibitors remarkably enhance the efficacy of the PARP inhibitor. Cotreatment with a DNMT inhibitor guadecitabine and a PARP inhibitor talazoparib effectively decreases xenograft tumor growth and increases overall survival in a BRCA-proficient high-grade serous ovarian cancer and TNBC models [78].

Epigenetic therapy combining with immunotherapy

Epigenetic alteration is recognized as one of the major mechanisms for cancers to evade the immunity and resist the immune-checkpoint blockage. Aberrant epigenetic progresses suppress antigen presentation, chemokine expression, T-cell infiltration, and immune cell cytotoxic effect, which cause poor immune response. However, such immune evasion can be reversed by targeting epigenetic progress using epigenetic therapy. Immune-checkpoint blockage exerts better suppression under the epigenetic inhibition. Combining epigenetic therapy and immunotherapy will provide a newly promising approach to kill cancer cells.

Combination of HDAC inhibitors and immunotherapy

The response to PD-1 blockades has been associated with T-cell infiltration, Th1/IFNγ signature, and tumor PD-L1 expression, suggesting that therapies augmenting T-cell infiltration or effector function may synergize with immune-checkpoint blockade [79]. It is found that HDAC inhibitors can increase expression of multiple T-cell chemokines in cancer cells, macrophages, and T cells, followed by enhanced T-cell infiltration and T-cell–dependent tumor regression. This augmented effect results from HDAC inhibition that impacts IFNγ-induced gene expression [80]. It is noteworthy that anti-PD-1 enhances T-cell chemokine expression and T-cell infiltration, which is mediated by IFNγ production as well [81]. A combination therapy of romidepsin with checkpoint blockade significantly enhances activation of tumor-infiltrating T cells by promoting a chemokine-IFNγ–positive feedback loop. In mice treated with romidepsin and anti-PD-1, tumor volume is substantially reduced by approximately 50% [82].

On the other hand, it is reported that HDAC inhibitors also can impact PD-L1 expression. Both human and murine cell lines and patient tumors treated with a variety of HDAC inhibitors in vitro show higher levels of PD-L1 and PD-L2. Due to HDAC inhibition, a more relaxed chromatin state is formed at the promoter regions of PD-L1 and PD-L2, inducing increased gene expression. The PD-L1 increase suggests a possibility that HDAC inhibition therapy may need to partner with PD-1 blockage. Combining a pan-HDAC inhibitor LBH589 with the PD-1 blockade synergistically reduces the tumor burden and improves the survival rate in a murine B16F10 model [83].

Immunosuppressive cells including a polymorphonuclear and monocytic myeloid derived suppressor cell (MDSC) can cause a poor immune response in cancers [84]. The mechanism of improving the immune response by HDAC inhibitors is complex. It is demonstrated that HDAC inhibitors may directly target STAT3 signaling to impair the T-cell–inhibiting activity of MDSCs, through downregulating immunosuppressive mediators including arginase 1, iNOS, and Cox-2. Liberation of immunosuppression by HDAC inhibition improves the anticancer effect of PD-1 antibodies. The combination of an HDAC inhibitor entinostat and anti-PD-1 treatment is very effective in enhancing tumor growth inhibition compared to the control and each single treatment group in RENCA and LLC syngeneic mouse cancer models [85].

Furthermore, a triple combination strategy with an HDAC inhibitor and PD-1 antibody and CTLA-4 antibody is hypothesized to cure immunogenic CT26 tumors or metastatic 4T1 tumors. It is demonstrated that such a triple combined therapy is more than sufficient to eradicate both primary tumors and metastasis in 4T1 tumor-bearing mice [86]. A triple combination strategy including anti-PD-1, anti-CTLA-4, and an HDAC inhibitor belinostat is also applied to treat HCC. In the Hepa129 tumor model, the triple combination induces tumor eradication in all treated mice and a 100% long-term survival, compared to a 60%–70% survival rate of mice treated with an anti-CTLA-4 plus belinostat combination or dual checkpoint blockades (anti-PD-1 and anti-CTLA-4) [87].

The loss of CD20 expression after the treatment of a CD20 antibody rituximab is recognized as the mechanism for rituximab resistance [88]. The rituximab-induced downregulation of CD20 is partly due to the histone deacetylation that prevents transcription factors from binding to DNA. As a result, inhibiting HDAC can upregulate CD20 expression and augment rituximab-induced cell death in malignant B cells. Cotreatment with an HDAC inhibitor chidamide and rituximab significantly induces cell death in vitro and substantially inhibits tumor growth in DLBCL-bearing xenograft mice [89].

Combination of BET inhibitors and immunotherapy

Recent works have clarified the relationship between BET function and PD-1/PD-L1 axis. Hogg et al. report that PD-L1 is a direct target of BET Inhibition. BET inhibitors decrease BRD4 occupancy at the PD-L1 locus, resulting in transcriptional inhibition of PD-L1. Cocombating the PD-1/PD-L1 axis, BET inhibitors can team up with anti-PD-1 inhibitors to improve an immune response. As a result, a combination therapy with JQ1 and an anti-PD1 Ab provides a more robust and sustainable response in mice bearing Eu-Myc lymphomas [90].

The suppression of the PD-1/PD-L1 axis by BET inhibition has also been reported in breast cancer cell models. BET proteins control PD-1 expression in T cells and PD-L1 in breast cancer cells partly through the mediating IFNγ pathway [91]. Based on the role of BET inhibitors in regulating the PD-1/PD-L1 axis, several clinical trials are under way in a combination with anti-PD-1 or anti-PD-L1 antibodies on cancer patients including TNBC patients (NCT03059147, NCT02419417, NCT03292172).

Since BET inhibition can improve anticancer immune response, it is used to combine with another typical immune-checkpoint inhibitor, anti-CTLA-4 inhibitor as well. Compared to anti-CTLA-4 alone, combined JQ1 and anti-CTLA-4 treatment induces a significant increase in the CD8/Treg ratio. In syngeneic murine Myc-Cap prostate cancer models, animals treated with a combination of JQ1 and anti-CTLA-4 show a 12.2% longer median survival rate than these treated with anti-CTLA-4 alone [92].

Immunomodulatory drugs (IMiDs) display a substantial antiproliferative effect against a majority of primary effusion lymphoma (PEL) cell lines, by arresting cell cycle at the G0/G1 phase. MYC depletion can further enhance the cytotoxicity of IMiDs [93]. It is well established that BET inhibition can suppress MYC transcription, providing a basis for a combination therapy with IMiDs. Indeed, an IMiD lenalidomide displays synergistic cytotoxicity with several structurally distinct BRD4 inhibitors (JQ1, IBET151, and PFI-1). Furthermore, combined administration of lenalidomide and JQ1 significantly increases the survival of PEL bearing NOD-SCID mice in an orthotopic xenograft model compared to each agent alone [94]. The similar combination strategy also works well in multiple myeloma in the study of Siu et al. [95].

Combination of EZH2 inhibitors and immunotherapy

Treatment with the anti-CTLA-4 antibody ipilimumab can increase EZH2 expression in human T cells, which inversely correlates with clinical outcome in a cohort of patients with prostate cancer. Obviously, EZH2 upregulation mediated by anti-CTLA-4 treatment in T cells modulates T-cell response and diminishes the effectiveness of anti-CTLA-4 therapy. EZH2 inhibition by CPI-1205 increases the effector-like T-cell response and enhances the effectiveness of anti-CTLA-4 therapy in tumor-bearing mice. The combination therapy of CPI-1205 and ipilimumab significantly reduces tumor growth and increases survival rate in both MB49 and B16-F10 cancer xenograft models as compared with anti-CTLA-4 monotherapy [96].

The treatment with the anti-CTLA-4 antibody induces the intratumoral tumor necrosis factor-α (TNF-α) production and T-cell accumulation, which also can increase EZH2 expression in cancer cells like melanoma. Furthermore, the cancer cells in turn silence their own immunogenicity and antigen presentation. Consistently, inhibition of EZH2 can reverse such immune therapy resistance, providing a basis for a combination strategy. Zingg et al. report that such a combination significantly delays tumor melanoma growth and reduces tumor volume by over 75% [97].

However, it is shown that EZH2 may play an opposite role in different immune cells. Specially, in RAW264.7 macrophages that can conduct a direct cytotoxic effect to mesothelioma cells by cell-to-cell contact, EZH2 inhibition surprisingly impairs such effect by inducing

PD-1 overexpression. RAW264.7 macrophages pretreated with the EZH2 inhibitor fail to control the tumor growth of AB1 and AB12 mesothelioma cells, which can be rescued by blocking PD-1 [98].

Combination of LSD1 inhibitors and immunotherapy

In recent preclinical studies, LSD1 inhibition is able to increase the expression of endogenous retroviral elements (ERVs) and decrease the expression of RNA-induced silencing complex (RISC) components. Subsequently, the events lead to endogenous double-stranded RNA (dsRNA) stress and type 1 interferon activation, which stimulate antitumor T-cell immunity and promote T-cell infiltration in low or noneimmunogenic tumors [99]. Abrogation of LSD1 by a high specific LSD1 inhibitor GSK-LSD1 in combination with a PD-1 blockade shows a striking effect on suppressing B16 tumor growth [100]. In this case, LSD1 inhibition overcomes the resistance of mouse melanoma to the PD-1 blockade, suggesting that LSD1 inhibition may be an effective adjuvant treatment with immunotherapy for poorly immunogenic tumors.

Additionally, Qin et al. demonstrate that LSD1 is involved in the downregulation of immune checkpoint regulators and effector T-cell chemokines in cancer, resulting in reduced effector T-cell trafficking to the tumor microenvironment. From the Cancer Genome Atlas (TCGA) data analysis, it is found that there is an inverse correlation between LSD1 expression and the levels of cytotoxic T cell-attracting chemokines and PD-L1 in clinical TNBC specimens. Therefore, LSD1 inhibition can reverse the aberrant suppression of effector T-cell chemokines and improve the infiltration of cytotoxic T cells, including CD8+ T cells. In mice bearing TNBC xenograft tumors, anti-PD-1 monoclonal antibody alone fails to elicit an obvious therapeutic effect. However, combining LSD1 inhibitors with the PD-1 antibody significantly suppresses tumor growth and pulmonary metastasis, with a 70% reduction in tumor burden compared to the vehicle control group [101]. Therefore, LSD1 inhibition can augment antitumor immune response and improve the therapeutic efficacy of immune-checkpoint blockade in cancers.

Maes et al. also show that cotreatment with the anti-PD-1 antibody and an LSD1 inhibitor ORY-1001 on the melanoma mice model has enhanced anticancer activity compared to the treatment with the anti-PD-1 antibody alone, corroborating the synergistic effects of the combination therapy [102]. Overall, these preclinical data suggest that combining LSD1 inhibitors with PD-1 blockades may be a promising therapeutic approach to treat cancer since it can overcome the common problem of immunotherapy resistance. Recently, this new combinatorial strategy is being tested in a phase I clinical trial assessing the safety and efficacy of an LSD1 inhibitor CC-90011 in combination with nivolumab, an anti-PD1 monoclonal antibody in patients with small cell lung cancer (ClinicalTrials.gov identifier: NCT04350463).

Conclusion and prospect

There have been tremendous advancements in the development of epigenetic drugs, and epigenetic therapy is often used in cancer treatment, especially for hematological cancers.

More evidences imply that a combination therapy of epigenetic therapy with other therapeutic methods substantially improves the curative efficacy. On one hand, either intrinsic or acquired drug resistance to the current therapies limits the drug efficiency. On the other hand, the intrinsic heterogenicity and plasticity features in cancer cells require multiple agents to control the malignancy. In future, better knowledge of the role and mechanisms of action of individual therapy, including epigenetic therapy, chemotherapy, and immunotherapy, is required to design a more precise combination therapy approach to target different cancers.

References

[1] Seebacher NA, Stacy AE, Porter GM, Merlot AM. Clinical development of targeted and immune based anti-cancer therapies. J Exp Clin Cancer Res 2019;38(1):156.

[2] Hoelder S, Clarke PA, Workman P. Discovery of small molecule cancer drugs: successes, challenges and opportunities. Mol Oncol 2012;6(2):155–76.

[3] Lee YT, Tan YJ, Oon CE. Molecular targeted therapy: treating cancer with specificity. Eur J Pharmacol 2018;834:188–96.

[4] Ward RA, Fawell S, Floc'h N, Flemington V, McKerrecher D, Smith PD. Challenges and opportunities in cancer drug resistance. Chem Rev 2021;121(6):3297–351.

[5] Vasan N, Baselga J, Hyman DM. A view on drug resistance in cancer. Nature 2019;575(7782):299–309.

[6] Leary M, Heerboth S, Lapinska K, Sarkar S. Sensitization of drug resistant cancer cells: a matter of combination therapy. Cancers (Basel) 2018;10(12).

[7] Whiteside TL, Demaria S, Rodriguez-Ruiz ME, Zarour HM, Melero I. Emerging opportunities and challenges in cancer immunotherapy. Clin Cancer Res 2016;22(8):1845–55.

[8] Vinay DS, Ryan EP, Pawelec G, Talib WH, Stagg J, Elkord E, et al. Immune evasion in cancer: mechanistic basis and therapeutic strategies. Semin Cancer Biol 2015;35(Suppl):S185–98.

[9] Gomez S, Tabernacki T, Kobyra J, Roberts P, Chiappinelli KB. Combining epigenetic and immune therapy to overcome cancer resistance. Semin Cancer Biol 2020;65:99–113.

[10] Jones PA, Ohtani H, Chakravarthy A, De Carvalho DD. Epigenetic therapy in immune-oncology. Nat Rev Cancer 2019;19(3):151–61.

[11] Yang Y. Cancer immunotherapy: harnessing the immune system to battle cancer. J Clin Invest 2015;125(9):3335–7.

[12] Keir ME, Butte MJ, Freeman GJ, Sharpe AH. PD-1 and its ligands in tolerance and immunity. Annu Rev Immunol 2008;26:677–704.

[13] Weiping Z, Jedd D, Wolchok, Lieping C. PD-L1 (B7-H1) and PD-1 pathway blockade for cancer therapy: mechanisms, response biomarkers, and combinations. Sci Transl Med 2016.

[14] Chen L, Han X. Anti-PD-1/PD-L1 therapy of human cancer: past, present, and future. J Clin Invest 2015;125(9):3384–91.

[15] Topalian SL, Drake CG, Pardoll DM. Immune checkpoint blockade: a common denominator approach to cancer therapy. Cancer Cell 2015;27(4):450–61.

[16] Sharma P, Hu-Lieskovan S, Wargo JA, Ribas A. Primary, adaptive, and acquired resistance to Cancer immunotherapy. Cell 2017;168(4):707–23.

[17] Patel SA, Minn AJ. Combination cancer therapy with immune checkpoint blockade: mechanisms and strategies. Immunity 2018;48(3):417–33.

[18] Jones PA, Issa JP, Baylin S. Targeting the cancer epigenome for therapy. Nat Rev Genet 2016;17(10):630–41.

[19] Ahuja N, Sharma AR, Baylin SB. Epigenetic therapeutics: a new weapon in the war against cancer. Annu Rev Med 2016;67(1):73–89.

[20] Morel D, Jeffery D, Aspeslagh S, Almouzni G, Postel-Vinay S. Combining epigenetic drugs with other therapies for solid tumours - past lessons and future promise. Nat Rev Clin Oncol 2020;17(2):91–107.

[21] Wilting RH, Dannenberg J-H. Epigenetic mechanisms in tumorigenesis, tumor cell heterogeneity and drug resistance. Drug Resist Updat 2012;15(1–2):21–38.

[22] Shah K, Rawal RM. Genetic and epigenetic modulation of drug resistance in Cancer: challenges and opportunities. Curr Drug Metab 2019;20(14):1114–31.

[23] Cao J, Yan Q. Cancer epigenetics, tumor immunity, and immunotherapy. Trends Cancer 2020;6(7):580–92.

[24] Topper MJ, Vaz M, Marrone KA, Brahmer JR, Baylin SB. The emerging role of epigenetic therapeutics in immuno-oncology. Nat Rev Clin Oncol 2020;17(2):75–90.

[25] Wang L, Liu X, Ren Y, Zhang J, Chen J, Zhou W, et al. Cisplatin-enriching cancer stem cells confer multidrug resistance in non-small cell lung cancer via enhancing TRIB1/HDAC activity. Cell Death Dis 2017;8(4), e2746.

[26] Toh TB, Lim JJ, Chow EK. Epigenetics in cancer stem cells. Mol Cancer 2017;16(1):29.

[27] Sun Y, Bao X, Ren Y, Jia L, Zou S, Han J, et al. Targeting HDAC/OAZ1 axis with a novel inhibitor effectively reverses cisplatin resistance in non-small cell lung cancer. Cell Death Dis 2019;10(6):400.

[28] Pitroda SP, Pashtan IM, Logan HL, Budke B, Darga TE, Weichselbaum RR, et al. DNA repair pathway gene expression score correlates with repair proficiency and tumor sensitivity to chemotherapy. Sci Transl Med 2014;6(229):229ra42.

[29] Powell SN, Kachnic LA. Roles of BRCA1 and BRCA2 in homologous recombination, DNA replication fidelity and the cellular response to ionizing radiation. Oncogene 2003;22(37):5784–91.

[30] Roos WP, Krumm A. The multifaceted influence of histone deacetylases on DNA damage signalling and DNA repair. Nucleic Acids Res 2016;44(21):10017–30.

[31] Yin L, Liu Y, Peng Y, Peng Y, Yu X, Gao Y, et al. PARP inhibitor veliparib and HDAC inhibitor SAHA synergistically co-target the UHRF1/BRCA1 DNA damage repair complex in prostate cancer cells. J Exp Clin Cancer Res 2018;37(1):153.

[32] Frederick BA, Helfrich BA, Coldren CD, Zheng D, Chan D, Bunn Jr PA, et al. Epithelial to mesenchymal transition predicts gefitinib resistance in cell lines of head and neck squamous cell carcinoma and non-small cell lung carcinoma. Mol Cancer Ther 2007;6(6):1683–91.

[33] Bruzzese F, Leone A, Rocco M, Carbone C, Piro G, Caraglia M, et al. HDAC inhibitor vorinostat enhances the antitumor effect of gefitinib in squamous cell carcinoma of head and neck by modulating ErbB receptor expression and reverting EMT. J Cell Physiol 2011;226(9):2378–90.

[34] Costa DB, Halmos B, Kumar A, Schumer ST, Huberman MS, Boggon TJ, et al. BIM mediates EGFR tyrosine kinase inhibitor-induced apoptosis in lung cancers with oncogenic EGFR mutations. PLoS Med 2007;4(10), e315.

[35] Xargay-Torrent S, Lopez-Guerra M, Saborit-Villarroya I, Rosich L, Campo E, Roue G, et al. Vorinostat-induced apoptosis in mantle cell lymphoma is mediated by acetylation of proapoptotic BH3-only gene promoters. Clin Cancer Res 2011;17(12):3956–68.

[36] Citro S, Bellini A, Miccolo C, Ghiani L, Carey TE, Chiocca S. Synergistic antitumour activity of HDAC inhibitor SAHA and EGFR inhibitor gefitinib in head and neck cancer: a key role for DeltaNp63alpha. Br J Cancer 2019;120(6):658–67.

[37] Tanimoto A, Takeuchi S, Arai S, Fukuda K, Yamada T, Roca X, et al. Histone deacetylase 3 inhibition overcomes BIM deletion polymorphism-mediated Osimertinib resistance in EGFR-mutant lung cancer. Clin Cancer Res 2017;23(12):3139–49.

[38] Yu W, Lu W, Chen G, Cheng F, Su H, Chen Y, et al. Inhibition of histone deacetylases sensitizes EGF receptor-TK inhibitor-resistant non-small-cell lung cancer cells to erlotinib in vitro and in vivo. Br J Pharmacol 2017; 174(20):3608–22.

[39] Huang Z, Zhou W, Li Y, Cao M, Wang T, Ma Y, et al. Novel hybrid molecule overcomes the limited response of solid tumours to HDAC inhibitors via suppressing JAK1-STAT3-BCL2 signalling. Theranostics 2018;8(18): 4995–5011.

[40] Gatla HR, Zou Y, Uddin MM, Singha B, Bu P, Vancura A, et al. Histone deacetylase (HDAC) inhibition induces IkappaB kinase (IKK)-dependent Interleukin-8/CXCL8 expression in ovarian Cancer cells. J Biol Chem 2017;292(12):5043–54.

[41] Deng C, Lipstein M, Rodriguez R, Serrano XO, McIntosh C, Tsai WY, et al. The novel IKK2 inhibitor LY2409881 potently synergizes with histone deacetylase inhibitors in preclinical models of lymphoma through the downregulation of NF-kappaB. Clin Cancer Res 2015;21(1):134–45.

[42] Stratikopoulos EE, Parsons RE. Molecular pathways: targeting the PI3K pathway in cancer-BET inhibitors to the rescue. Clin Cancer Res 2016;22(11):2605–10.

[43] Zhan F, Huang Y, Colla S, Stewart JP, Hanamura I, Gupta S, et al. The molecular classification of multiple myeloma. Blood 2006;108(6):2020–8.

[44] Stubbs MC, Burn TC, Sparks R, Maduskuie T, Diamond S, Rupar M, et al. The novel bromodomain and extraterminal domain inhibitor INCB054329 induces vulnerabilities in myeloma cells that inform rational combination strategies. Clin Cancer Res 2019;25(1):300–11.

[45] Saenz DT, Fiskus W, Manshouri T, Rajapakshe K, Krieger S, Sun B, et al. BET protein bromodomain inhibitor-based combinations are highly active against post-myeloproliferative neoplasm secondary AML cells. Leukemia 2017;31(3):678–87.

[46] Grivennikov SI, Karin M. Dangerous liaisons: STAT3 and NF-kappaB collaboration and crosstalk in cancer. Cytokine Growth Factor Rev 2010;21(1):11–9.

[47] Kleppe M, Koche R, Zou L, van Galen P, Hill CE, Dong L, et al. Dual targeting of oncogenic activation and inflammatory signaling increases therapeutic efficacy in myeloproliferative neoplasms. Cancer Cell 2018; 33(1):29–43 [e7].

[48] Sun B, Shah B, Fiskus W, Qi J, Rajapakshe K, Coarfa C, et al. Synergistic activity of BET protein antagonist-based combinations in mantle cell lymphoma cells sensitive or resistant to ibrutinib. Blood 2015;126(13):1565–74.

[49] Ceribelli M, Kelly PN, Shaffer AL, Wright GW, Xiao W, Yang Y, et al. Blockade of oncogenic IkappaB kinase activity in diffuse large B-cell lymphoma by bromodomain and extraterminal domain protein inhibitors. Proc Natl Acad Sci U S A 2014;111(31):11365–70.

[50] Sun CY, Nie J, Huang JP, Zheng GJ, Feng B. Targeting STAT3 inhibition to reverse cisplatin resistance. Biomed Pharmacother 2019;117:109135.

[51] Bagratuni T, Mavrianou N, Gavalas NG, Tzannis K, Arapinis C, Liontos M, et al. JQ1 inhibits tumour growth in combination with cisplatin and suppresses JAK/STAT signalling pathway in ovarian cancer. Eur J Cancer 2020;126:125–35.

[52] Yecies D, Carlson NE, Deng J, Letai A. Acquired resistance to ABT-737 in lymphoma cells that up-regulate MCL-1 and BFL-1. Blood 2010;115(16):3304–13.

[53] Conery AR, Centore RC, Spillane KL, Follmer NE, Bommi-Reddy A, Hatton C, et al. Preclinical anticancer efficacy of BET Bromodomain inhibitors is determined by the apoptotic response. Cancer Res 2016;76 (6):1313–9.

[54] Peirs S, Frismantas V, Matthijssens F, Van Loocke W, Pieters T, Vandamme N, et al. Targeting BET proteins improves the therapeutic efficacy of BCL-2 inhibition in T-cell acute lymphoblastic leukemia. Leukemia 2017;31(10):2037–47.

[55] Esteve-Arenys A, Valero JG, Chamorro-Jorganes A, Gonzalez D, Rodriguez V, Dlouhy I, et al. The BET bromodomain inhibitor CPI203 overcomes resistance to ABT-199 (venetoclax) by downregulation of BFL-1/A1 in in vitro and in vivo models of MYC+/BCL2+ double hit lymphoma. Oncogene 2018;37(14):1830–44.

[56] Bechter O, Schoffski P. Make your best BET: the emerging role of BET inhibitor treatment in malignant tumors. Pharmacol Ther 2020;208:107479.

[57] Shu S, Wu HJ, Ge JY, Zeid R, Harris IS, Jovanovic B, et al. Synthetic lethal and resistance interactions with BET Bromodomain inhibitors in triple-negative breast Cancer. Mol Cell 2020;78(6):1096–113. e8.

[58] Jin X, Yan Y, Wang D, Ding D, Ma T, Ye Z, et al. DUB3 promotes BET inhibitor resistance and Cancer progression by deubiquitinating BRD4. Mol Cell 2018;71(4):592–605. e4.

[59] Sun J, Cai X, Yung MM, Zhou W, Li J, Zhang Y, et al. miR-137 mediates the functional link between c-Myc and EZH2 that regulates cisplatin resistance in ovarian cancer. Oncogene 2019;38(4):564–80.

[60] Deb G, Thakur VS, Gupta S. Multifaceted role of EZH2 in breast and prostate tumorigenesis: epigenetics and beyond. Epigenetics 2013;8(5):464–76.

[61] Nie L, Wei Y, Zhang F, Hsu YH, Chan LC, Xia W, et al. CDK2-mediated site-specific phosphorylation of EZH2 drives and maintains triple-negative breast cancer. Nat Commun 2019;10(1):5114.

[62] Watson PA, Arora VK, Sawyers CL. Emerging mechanisms of resistance to androgen receptor inhibitors in prostate cancer. Nat Rev Cancer 2015;15(12):701–11.

[63] Kim J, Lee Y, Lu X, Song B, Fong K-W, Cao Q, et al. Polycomb- and methylation-independent roles of EZH2 as a transcription activator. Cell Rep 2018;25(10):2808–20. e4.

[64] Yamaguchi H, Du Y, Nakai K, Ding M, Chang SS, Hsu JL, et al. EZH2 contributes to the response to PARP inhibitors through its PARP-mediated poly-ADP ribosylation in breast cancer. Oncogene 2018;37(2): 208–17.

[65] St Pierre R, Kadoch C. Mammalian SWI/SNF complexes in cancer: emerging therapeutic opportunities. Curr Opin Genet Dev 2017;42:56–67.

[66] Wu S, Fatkhutdinov N, Fukumoto T, Bitler BG, Park PH, Kossenkov AV, et al. SWI/SNF catalytic subunits' switch drives resistance to EZH2 inhibitors in ARID1A-mutated cells. Nat Commun 2018;9(1):4116.

[67] Verigos J, Karakaidos P, Kordias D, Papoudou-Bai A, Evangelou Z, Harissis HV, et al. The histone demethylase LSD1/KDM1A mediates chemoresistance in breast cancer via regulation of a stem cell program. Cancers 2019;11(10).

[68] Peng W, Zhang H, Tan S, Li Y, Zhou Y, Wang L, et al. Synergistic antitumor effect of 5-fluorouracil with the novel LSD1 inhibitor ZY0511 in colorectal cancer. Ther Adv Med Oncol 2020;12. 1758835920937428.

[69] Schenk T, Chen WC, Gollner S, Howell L, Jin L, Hebestreit K, et al. Inhibition of the LSD1 (KDM1A) demethylase reactivates the all-trans-retinoic acid differentiation pathway in acute myeloid leukemia. Nat Med 2012;18(4):605–11.

[70] Smitheman KN, Severson TM, Rajapurkar SR, McCabe MT, Karpinich N, Foley J, et al. Lysine specific demethylase 1 inactivation enhances differentiation and promotes cytotoxic response when combined with all-trans retinoic acid in acute myeloid leukemia across subtypes. Haematologica 2019;104(6):1156–67.

[71] Watts JM, Bradley TJ, Thomassen A, Dinh YT, Tejera D, Englund K, et al. The lysine-specific demethylase 1 (LSD1) inhibitor tranylcypromine (TCP) in combination with ATRA is tolerable and has anti-leukemic activity in adult patients with relapsed/refractory AML and MDS. Blood 2018;132(Suppl 1):2721.

[72] Deb G, Wingelhofer B, Amaral FMR, Maiques-Diaz A, Chadwick JA, Spencer GJ, et al. Pre-clinical activity of combined LSD1 and mTORC1 inhibition in MLL-translocated acute myeloid leukaemia. Leukemia 2020; 34(5):1266–77.

[73] Abdel-Aziz AK, Pallavicini I, Ceccacci E, Meroni G, Saadeldin MK, Varasi M, et al. Tuning mTORC1 activity dictates the response of acute myeloid leukemia to LSD1 inhibition. Haematologica 2020;105(8):2105–17.

[74] Lu Y, Liu Y, Oeck S, Glazer PM. Hypoxia promotes resistance to EGFR inhibition in NSCLC cells via the histone demethylases, LSD1 and PLU-1. Mol Cancer Res 2018;16(10):1458–69.

[75] Huang M, Chen C, Geng J, Han D, Wang T, Xie T, et al. Targeting KDM1A attenuates Wnt/beta-catenin signaling pathway to eliminate sorafenib-resistant stem-like cells in hepatocellular carcinoma. Cancer Lett 2017;398:12–21.

[76] Wu LW, Zhou DM, Zhang ZY, Zhang JK, Zhu HJ, Lin NM, et al. Suppression of LSD1 enhances the cytotoxic and apoptotic effects of regorafenib in hepatocellular carcinoma cells. Biochem Biophys Res Commun 2019; 512(4):852–8.

[77] Caiado F, Maia-Silva D, Jardim C, Schmolka N, Carvalho T, Reforco C, et al. Lineage tracing of acute myeloid leukemia reveals the impact of hypomethylating agents on chemoresistance selection. Nat Commun 2019; 10(1):4986.

[78] Pulliam N, Fang F, Ozes AR, Tang J, Adewuyi A, Keer H, et al. An effective epigenetic-PARP inhibitor combination therapy for breast and ovarian cancers independent of BRCA mutations. Clin Cancer Res 2018; 24(13):3163–75.

[79] Herbst RS, Soria JC, Kowanetz M, Fine GD, Hamid O, Gordon MS, et al. Predictive correlates of response to the anti-PD-L1 antibody MPDL3280A in cancer patients. Nature 2014;515(7528):563–7.

[80] Sarkar D, Leung EY, Baguley BC, Finlay GJ, Askarian-Amiri ME. Epigenetic regulation in human melanoma: past and future. Epigenetics 2015;10(2):103–21.

[81] Peng W, Liu C, Xu C, Lou Y, Chen J, Yang Y, et al. PD-1 blockade enhances T-cell migration to tumors by elevating IFN-gamma inducible chemokines. Cancer Res 2012;72(20):5209–18.

[82] Zheng H, Zhao W, Yan C, Watson CC, Massengill M, Xie M, et al. HDAC inhibitors enhance T-cell chemokine expression and augment response to PD-1 immunotherapy in lung adenocarcinoma. Clin Cancer Res 2016; 22(16):4119–32.

[83] Woods DM, Sodre AL, Villagra A, Sarnaik A, Sotomayor EM, Weber J. HDAC inhibition upregulates PD-1 ligands in melanoma and augments immunotherapy with PD-1 blockade. Cancer Immunol Res 2015; 3(12):1375–85.

[84] Gabrilovich DI. Myeloid-derived suppressor cells. Cancer Immunol Res 2017;5(1):3–8.

[85] Orillion A, Hashimoto A, Damayanti N, Shen L, Adelaiye-Ogala R, Arisa S, et al. Entinostat neutralizes myeloid-derived suppressor cells and enhances the antitumor effect of PD-1 inhibition in murine models of lung and renal cell carcinoma. Clin Cancer Res 2017;23(17):5187–201.

[86] Kim K, Skora AD, Li Z, Liu Q, Tam AJ, Blosser RL, et al. Eradication of metastatic mouse cancers resistant to immune checkpoint blockade by suppression of myeloid-derived cells. Proc Natl Acad Sci U S A 2014; 111(32):11774–9.

[87] Llopiz D, Ruiz M, Villanueva L, Iglesias T, Silva L, Egea J, et al. Enhanced anti-tumor efficacy of checkpoint inhibitors in combination with the histone deacetylase inhibitor Belinostat in a murine hepatocellular carcinoma model. Cancer Immunol Immunother 2019;68(3):379–93.

[88] Bobrowicz M, Dwojak M, Pyrzynska B, Stachura J, Muchowicz A, Berthel E, et al. HDAC6 inhibition upregulates CD20 levels and increases the efficacy of anti-CD20 monoclonal antibodies. Blood 2017; 130(14):1628–38.

[89] Guan XW, Wang HQ, Ban WW, Chang Z, Chen HZ, Jia L, et al. Novel HDAC inhibitor Chidamide synergizes with rituximab to inhibit diffuse large B-cell lymphoma tumour growth by upregulating CD20. Cell Death Dis 2020;11(1):20.

[90] Hogg SJ, Vervoort SJ, Deswal S, Ott CJ, Li J, Cluse LA, et al. BET-bromodomain inhibitors engage the host immune system and regulate expression of the immune checkpoint ligand PD-L1. Cell Rep 2017;18(9):2162–74.

[91] Andrieu GP, Shafran JS, Smith CL, Belkina AC, Casey AN, Jafari N, et al. BET protein targeting suppresses the PD-1/PD-L1 pathway in triple-negative breast cancer and elicits anti-tumor immune response. Cancer Lett 2019;465:45–58.

[92] Mao W, Ghasemzadeh A, Freeman ZT, Obradovic A, Chaimowitz MG, Nirschl TR, et al. Immunogenicity of prostate cancer is augmented by BET bromodomain inhibition. J Immunother Cancer 2019;7(1):277.

[93] Diaz T, Rodriguez V, Lozano E, Mena MP, Calderon M, Rosinol L, et al. The BET bromodomain inhibitor CPI203 improves lenalidomide and dexamethasone activity in in vitro and in vivo models of multiple myeloma by blockade of Ikaros and MYC signaling. Haematologica 2017;102(10):1776–84.

[94] Gopalakrishnan R, Matta H, Tolani B, Triche Jr T, Chaudhary PM. Immunomodulatory drugs target IKZF1-IRF4-MYC axis in primary effusion lymphoma in a cereblon-dependent manner and display synergistic cytotoxicity with BRD4 inhibitors. Oncogene 2016;35(14):1797–810.

[95] Siu KT, Ramachandran J, Yee AJ, Eda H, Santo L, Panaroni C, et al. Preclinical activity of CPI-0610, a novel small-molecule bromodomain and extra-terminal protein inhibitor in the therapy of multiple myeloma. Leukemia 2016;31(8):1760–9.

[96] Goswami S, Apostolou I, Zhang J, Skepner J, Anandhan S, Zhang X, et al. Modulation of EZH2 expression in T cells improves efficacy of anti-CTLA-4 therapy. J Clin Invest 2018;128(9):3813–8.

[97] Zingg D, Arenas-Ramirez N, Sahin D, Rosalia RA, Antunes AT, Haeusel J, et al. The histone methyltransferase Ezh2 controls mechanisms of adaptive resistance to tumor immunotherapy. Cell Rep 2017;20(4):854–67.

[98] Hamaidia M, Gazon H, Hoyos C, Hoffmann GB, Louis R, Duysinx B, et al. Inhibition of EZH2 methyltransferase decreases immunoediting of mesothelioma cells by autologous macrophages through a PD-1-dependent mechanism. JCI Insight 2019;4(18).

[99] Herbert A. ADAR and immune silencing in cancer. Trends Cancer 2019;5(5):272–82.

[100] Sheng W, LaFleur MW, Nguyen TH, Chen S, Chakravarthy A, Conway JR, et al. LSD1 ablation stimulates antitumor immunity and enables checkpoint blockade. Cell 2018;174(3):549–63. e19.

[101] Qin Y, Vasilatos SN, Chen L, Wu H, Cao Z, Fu Y, et al. Inhibition of histone lysine-specific demethylase 1 elicits breast tumor immunity and enhances antitumor efficacy of immune checkpoint blockade. Oncogene 2019; 38(3):390–405.

[102] Maes T, Mascaró C, Sacilotto N, Lufino MM, Ciceri F. Targeting KDM1A with iadademstat in combination with immunotherapy in an in vivo model of melanoma. J Clin Oncol 2019;37(15_suppl):e14248.

CHAPTER

8

Future epigenetic treatment strategies

Lingyue Gao and Lihui Wang

Department of Pharmacology, Shenyang Pharmaceutical University, Shenyang, People's Republic of China

Abstract

Until now, chemoresistance is still the main obstacle for clinical cancer management, which severely influences the prognosis of patients. Recently, given the essential role of epigenetic regulation during chemoresistance and the great progression of epigenetic therapy, researchers are now paying more attention to combination therapy involving epigenetic agents to overcome the issue. Encouragingly, several epigenetic agents have shown excellent effects in restoring the sensitivity of cancer cells to cytotoxic drugs both in laboratory and clinical studies. In the previous chapters, a new scenario based upon pharmacological weapons targeting the epigenetic regulation to overcoming chemoresistance was drawn. In chapter, we mainly discuss the epigenetic therapy acting on noncoding RNA to combat chemoresistance, the promise of predictive biomarkers in foretelling the therapy index, and the promising combination strategy combining epigenetic agents with targeted therapy.

Abbreviations

AML	acute myeloid leukemia
BET	bromodomain and extraterminal
ceRNA	competing endogenous RNA
DNMTi	DNA methyltransferase inhibitors
EGFR-TKIs	epidermal growth factor receptor-tyrosine kinase inhibitors
EMT	epithelial-mesenchymal transition
EZH2	enhancer of zeste homolog 2
HDACi	HDAC inhibiotrs
HIF1α	hypoxia-inducible factor 1α
HMT	histone lysine methyltransferases
IDH1	isocitrate dehydrogenase 1
LFUS	low-frequency ultrasound
lncRNA	long noncoding
miRNA	microRNA
ncRNA	noncoding RNA
NSCLC	non-small-cell lung cancer

Epigenetic Regulation in Overcoming Chemoresistance
https://doi.org/10.1016/B978-0-12-823684-0.00015-3

OS overall survival
PFS progression-free survival
TCM traditional Chinese medicine

Conflicts of interest

No potential conflicts of interest were disclosed.

Introduction

The promising epigenetic therapy

Remarkable progress in cancer treatment has been achieved over recent decades; yet chemoresistance is still one of the main issues for clinical cancer management nowadays. Since its concept was proposed by C.H. Waddington in the mid-20th century, considerable findings have confirmed that epigenetic regulation was closely associated with multiple aspects of tumor progression such as tumor microenvironment regulation and cancer stem cell formation, which were well-known factors that led to chemotherapy failure [1, 2]. For example, our previous work revealed that inhibition of histone deacetylase (HDAC) could enhance the antitumor effects of cisplatin in non-small-cell lung cancer (NSCLC), which was correlated with the reduction of cancer stem cell formation [3].

Moreover, unlike the genetic changes, epigenetic regulation can feasibly be reprogrammed [2]. This along with its pivotal role in chemoresistance makes epigenetic regulation a promising target to overcome therapy failure. Encouragingly, some achievements were caught recently discovered. There are already some epigenetic agents used to currently treat cancer in clinics (Table 1). Most important among them are HDAC inhibitors (HDACi) and DNA methyltransferase inhibitors (DNMTi); these drugs are primarily applied to hematological neoplasms [4–6]. This may partly be due to the reason that hematological neoplasms are more likely to have mutations and thus are more dependent on epigenetic regulation than solid cancers. Yet, solid cancers that have arisen from more differentiated cells may be another reason [7].

Inspired by these clinical advancements, more attention was paid to other epigenetic regulators such as histone lysine methyltransferases (HMT) in related studies. More endeavors were also made to treat solid cancers via epigenetic agents. The Tazemetostat is a representative example. As an inhibitor of enhancer of zeste homolog 2 (EZH2), it received approval for treatment of epithelioid sarcoma for people older than 16 years in the U.S. [8]. Also, as an indispensable part of epigenetic regulation, noncoding RNA (ncRNA) has also attracted attention from researchers currently pursuing in the field of cancer. The most fully investigated ncRNA are microRNA (miRNA) and long noncoding RNA (lncRNA) wth lengths of 20–25 nt and 200–100,000 nt, respectively. Their roles in chemoresistance and as pharmacological weapons will be discussed at length in later sections.

New approaches for epigenetic therapy

Although some remarkable progression has been achieved, it is worth noting that there are still some disappointing outcomes with epigenetic therapy especially in clinical trials. Faced

TABLE 1 Current epigenetic agents used in clinics.

Inhibitor	Drug	Target	Therapeutic strategy	Cancer type	Phase
DNMTi	Azacitidine	DNMT1, DNMT3A, DNMT3B	Monotherapy	AML, MDS	EMA and FDA approved
	Decitabine	DNMT1, DNMT3A, DNMT3B	Monotherapy	AML, MDS	EMA (for AML) FDA (for MDS) approved
HDACi	Belinostat	HDAC1, HDAC2	Monotherapy	PTCL;HCC, DLBCL, MDS, Burkitt lymphoma, thymic carcinoma	FDA approved
	Romidepsin	HDAC1	Monotherapy	CTCL, PTCL	FDA approved
	Panobinostat	HDAC1, HDAC2, HDAC4	Monotherapy	MM; thyroid carcinoma, RCC, breast cancer, AML	FDA approved
	Vorinostat	HDAC1, HDAC2, HDAC4	Monotherapy	CTCL	FDA approved
HMTi	Tazemetostat	EZH2	Monotherapy	Epithelioid sarcoma	FDA approved
IDHi	Idhifa	IDH, IDH2	Monotherapy	AML	FDA approved
	Enasidenib	IDH2	Monotherapy	PTCL	CFDA approved
	Tibsovo	IDH, IDH2	Monotherapy	R/R AML, Cholangiocarcinoma	FDA approved
ncRNA	Patisiran	miRNA	Monotherapy	Rare polyneuropathy	FDA approved

DNMTi, DNA methyltransferase inhibitors; *HDACi*, histone deacetylase inhibitors; *HMTi*, histone lysine methyltransferases inhibitors; *IDHi*, isocitrate dehydrogenase inhibitors; *ncRNA*, noncoding RNA; *AML*, acute myelocytic leukemia; *MDS*, myelodysplastic syndrome; *PTCL*, peripheral t cell lymphoma; *HCC*, hepatocellular carcinoma; *DLBCL*, diffuse large B-cell lymphoma; *MM*, multiple myeloma; *RCC*, renal cell carcinoma; *CTCL*, cutaneous t-cell lymphoma; *EMA*, European Medicines Agency; *FDA*, Food and Drug Administration; *CFDA*, China Food and Drug Administration.

with these issues, some new approaches may be needed. First, toxicity was a common issue both in monotherapy and combination therapy. Taking the following phase II trial as an example, a combination with HDACi vorinosant indeed improved the anticancer effects of carboplatin-paclitaxel therapy, which was represented by longer progression-free survival (PFS) and higher response rate, but the increased toxicity cannot be ignored [9]. Inspired by one phase I clinical trial that the DNMTi decitabine could effectively regulate the methylation level of CpG island at a moderate dose compared with the maximal tolerated dose, lower dosage of epigenetic agents in combination therapy was proposed [10]. Following this strategy, the toxicity of combination therapy may be effectively limited to some extent without influencing the epigenetic regulation effects.

Apart from lowering the dose, prolonged treatment duration may be another effective approach. This arose from observations that the isocitrate dehydrogenase 1 (IDH1) inhibitor ivosidenib exhibited a delayed action pattern in an AML clinical trial. It took several months to obtain the drug responses in the study [11]. Similar results were observed in a trial about myelodysplasia and AML. Its data indicated that patients need to spend several months to benefit from the administration of DNMTi 5-azacytidine [12]. These phenomena may be partly due to the adequate time needed for epigenetic reprograming to be obvious and effective. Given these findings, the overall survival (OS), the drug responses of patients following therapy, as well as the epigenetic changes of the cancer cells should be taken into account when evaluating the outcomes of epigenetic therapy in future.

What's more, a rational treatment schedule was also pursued, as the sequence of administration may affect the anticancer effects in combination therapy. For example, Diasio's group [13] carried out a phase I trial to determine a better treatment schedule of combination therapy involving vorinostat, Leucovorin, and 5-Fu for refractory colorectal cancer treatment. Furthermore, thorough clarifying of the role of epigenetic regulator in chemoresistance by the genome editing approach was recommended, which may provide a rationale for epigenetic drug design and combination therapy involving epigenetic agents. Last but not least, enlightened by targeted therapy, the merits of which is precision treatment, researchers are now exploring the predictive biomarkers for epigenetic therapy with hope to improve patients' benefits [12].

Briefly, given the achievements and merits of epigenetic therapy, it may have great potential when combined with chemotherapy, which to some extent has been approved in numerous preclinical and clinical trials. The current chapter emphasizes ncRNA as a target to reverse chemoresistance. The predictive biomarkers for precise epigenetic therapy are also outlined. Additionally, the combination therapy involving epigenetic agents plus targeted drugs are also discussed. We truly hope our work may inspire new ideas about the application of epigenetic therapy, especially in overcoming chemoresistance.

The therapies targeting ncRNA

miRNA in chemoresistance

Recently, accumulating evidence has verified the indispensable role of miRNA during cancer pathogenesis involving tumor metastasis, phenotypic transformation, chemoresistance, and so on [14]. The miRNA achieved its role mainly through complementarily binding to the 3′-untranslated region of its targeting mRNA to downregulate the target genes and silence them (Fig. 1) [15].

With regard to chemoresistance, miRNA was reported to regulate the therapy responses of cancer cells to different sorts of cytotoxic agents. For example, miR-223 downregulated PAX6, which then induced chemoresistance of glioblastoma to temozolomide [16]. Similarly, in breast cancer cells, miR-181b was found to inhibit its target Bim to regulate the caspase-dependent apoptosis and finally decreased sensitivity to doxorubicin treatment [17]. In addition to the oncogenic role, miRNA can also reverse chemoresistance.

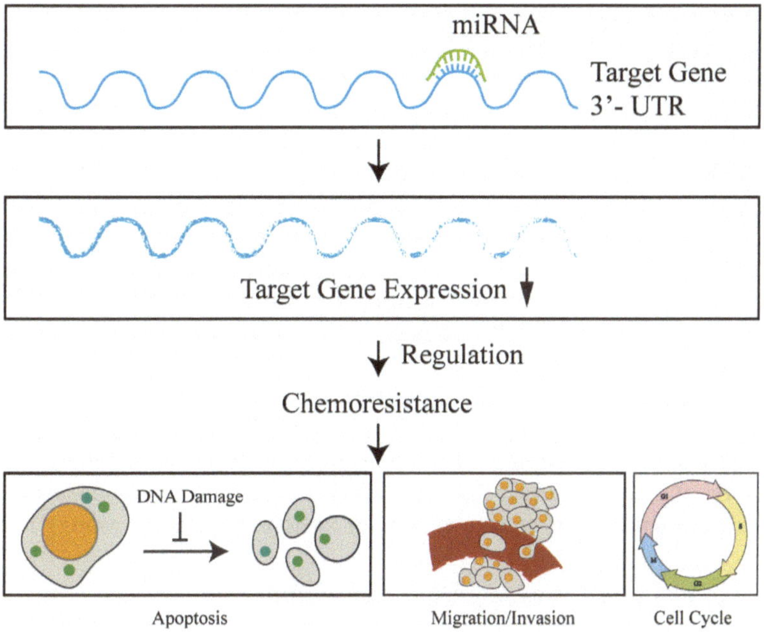

FIG. 1 The miRNA regulates the chemoresistance via targeting downstream genes.

An obvious example is that miR-4262 could target the PTEN/AKT pathway to affect cell migration, proliferation, and apoptosis, and then reverse the chemoresistance of lung cancer to paclitaxel both in vitro and in vivo [18].

These findings revealed the possibilities of reversing the failure of chemotherapy by targeting miRNA. But it was noteworthy that different miRNA may play opposite roles in the chemoresistance process. Even the same miRNA may exhibit opposite effects under different conditions. Take miR-223 as an example. As mentioned, it displayed an oncogenic role in temozolomide-resistant glioblastoma stem cells. Nevertheless, in hepatocellular carcinoma cells, miR-223 worked as a tumor suppressor, which enhanced the efficiency of doxorubicin by targeting the FOXO3a to attenuate the autophagy [19]. This intriguing phenomenon may due to differences in the cell model and drug treatment as the culture condition and the tumor microenvironment were key influential factors for epigenetic regulators [20]. Thus, further research is still needed to illustrate the exact role of different miRNA in cancer pathogenesis including chemoresistance.

Epigenetic therapies targeting miRNA

Nanostructured carriers

Given the essential of role of miRNA in chemoresistance, approaches affecting the miRNA were conducted with hope to overcome this hindrance of cancer management. Owing to their properties of high safety, feasible to modify, and high transfection efficiency, nanostructured carriers were employed [21]. These carriers mainly belong to three categories: lipid-based, cationic polymer-based, and exosomes [22].

As expected, notable progression was acquired in delivering miRNA or its siRNA through nanostructured carriers. A study by Liu's group [23] indicated that DOTAP complexed with miR-200c was effectively delivered into breast cancer stem cells. The DOTAP/miR-200c complexes downregulated miR-200's target class III β-tubulin and then restored the cytotoxic effects of paclitaxel. Furthermore, Xie et al. [24] structured a nanoparticle using PCX and conjugated it with anti-miR-210 to enhance the anticancer effects of combination treatment of gemcitabine and cisplatin by promoting cell apoptosis while inhibiting cell proliferation. They also found that intravenous administration with the nanocomplexes could reverse the resistance of stem cells to the combination therapy. In Liang's study [25], cell-derived exosomes were employed to deliver the 5-fluorouraci and miR-21 inhibitor to colon cancer cells. It could restore the sensitivity of cancer cells to chemotherapy through downregulating PTEN and hMSH2, which are the targets of miR-21.

These approaches offer a rationale of overcoming chemoresistance via targeting miRNA. With the advance of nanomaterials and in-depth understanding of the role of miRNA in chemotherapy, these trials may turn into clinical options for cancer management someday.

Bioactive compounds and Chinese herbal compounds

For a long time, intolerable adverse drug reactions and tumor heterogeneity are the main obstacles for clinical cancer management. In view of its properties of high safety and multiple targets, traditional Chinese medicine (TCM) has been a hotspot of cancer studies [26]. Recently, researchers have found that TCM and bioactive compounds isolated from it have the ability to reverse chemoresistance by affecting miRNA.

A study from Zeng et al. [27] demonstrated that curcumol, a bioactive component extracted from Rhizoma Curcumae, could upregulate miR-2-3p to reduce the level of drug transport ABBCC3 and thus promote drug accumulation. Through this machinery, curcumol enhanced drug response to doxorubicin in vivo and in vitro. Rosmarinic acid, another kind of TCM, has been verified to enhance therapy efficacy of 5-Fu by upregulating miR-6785/FOXO4 signaling pathway [28]. By targeting this axis, Rosmarinic acid could promote cell apoptosis by increasing Bax level and facilitate drug accumulation by decreasing p-glycoprotein level.

Just like the bioactive compounds, Chinese herbal compounds also exhibited effects of reversing chemoresistance via targeting miRNA. A notable example is that docetaxel displayed enhanced anticancer effects during treatment of prostate cancer with Fuzheng Yiliu decoction [29]. This phenomenon is mainly due to the Fuzheng Yiliu decoction that could target multiple miRNAs including miR-34b and miR-674-3p to regulate the prosurvival PI3K/AKT pathway.

These excellent findings may provide more inspiration for using TCM to reverse chemoresistance, especially with deeper understanding about the role of miRNA in cancer progression.

The lncRNA in chemoresistance

Another sort of ncRNA, lncRNA also has been confirmed to be involved in various biological properties of cancer including chemoresistance [30, 31]. Besides, lncRNA may regulate

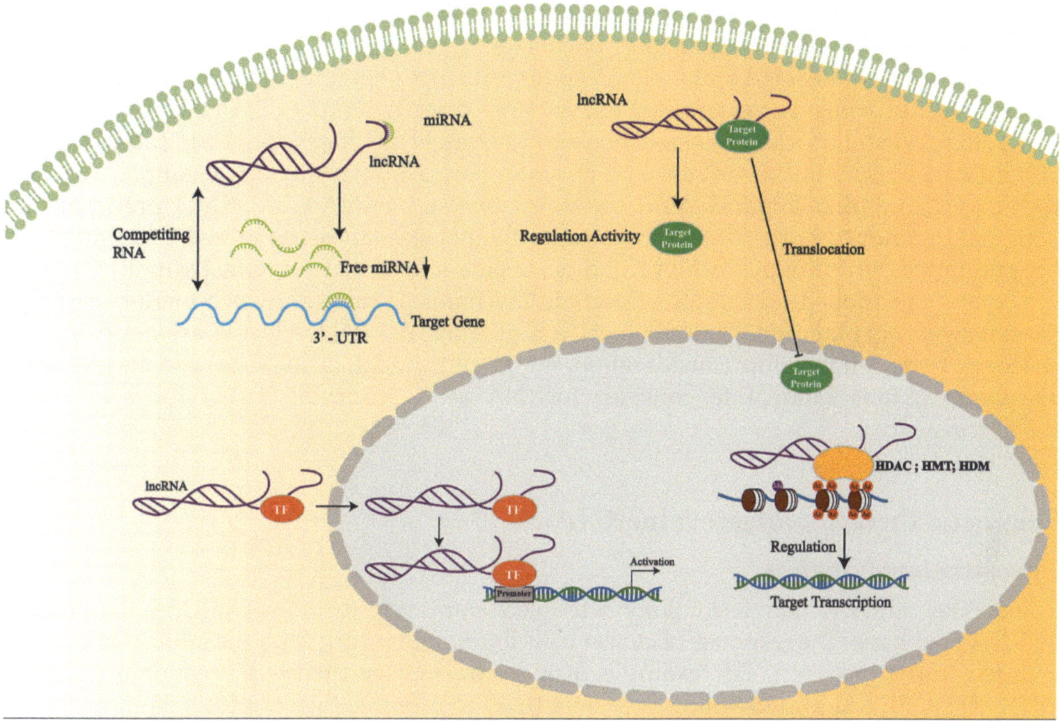

FIG. 2 The mechanisms through which lncRNA exerts its role. TF, transcription factor; HDAC, histone deacetylase; HMT, histone methyltransferase; HDM, histone demethylase.

this process through a variety of approaches (Fig. 2). One of the most known and well-investigated machinery is the competing endogenous RNA (ceRNA) pattern, as shown in the following examples. In ovarian cancer, lncRNA UCA1 has been verified to induce chemoresistance to cisplatin as a ceRNA of miR-143 [32]. During this process, lncRNA UCA1 targets the miR-143, which hampered the inhibitory effects of miR-143 and promoted the role of FOSL2, the target gene of miR-143. Similarly, in pancreatic cancer, lncRNA GAS5 targeted miR-221 to upregulate the suppressor of cytokine signaling 3 and finally reversed the gemcitabine resistance mediated by epithelial-mesenchymal transition (EMT) and stemness transition [33].

Moreover, it has been demonstrated that lncRNA could synergize with other epigenetic regulators such as EZH2 to regulate gene transcription. For example, lncRNA HOXD-AS1 recruited EZH2 to PDCD4 promoter and then increased its H3K27me3 level to inhibit transcription and expression of PDCD4 [34]. This process finally decreased the chemosensitivity of gastric cancer to cisplatin.

In addition to cooperating with epigenetic regulators to modulate gene transcription, lncRNA was also confirmed to regulate this process via influencing the transcription factors. In doxorubicin and paclitaxel-resistant breast cancer, LINC00160 accelerated the recruitment of transcription factor C/EBPβ to the TFF3 promoter region to induce its transcription and increase its expression [35]. This process could then enhance the drug responses of cancer cells when treated with doxorubicin or paclitaxel.

It has also been indicated that lncRNA could influence chemoresistance by directly affecting its target proteins. For example, lncRNA XIST directly binds to SMAD2, the effector of TGF-β, to prevent its nuclear translocation and ultimately facilitates apoptosis and pyroptosis of lung cancer cells in the presence of cisplatin [36]. Another obvious example is during chemoresistance of gallbladder cancer to doxoriubicin, lncRNA GBCDRlnc1 prevented the initiation of autophagy by directly binding to the phosphoglycerate kinase 1 to impede its ubiquitination, which finally achieved an oncogenic role of lncRNA GBCDRlnc1 [37].

The aforementioned findings are the main mechanisms through which lncRNA regulate chemoresistance; other mechanisms such as regulating the RNA splicing are not discussed at length here. Taking into consideration the large number of lncRNA and its extensive mechanisms, it is rational to consider lncRNA as a promising target for overcoming chemoresistance.

Epigenetic therapies targeting lncRNA

Nanostructured carriers

Just like miRNA, nanotechnology was also employed to deliver siRNA that targets lncRNA to enhance the response of cancer cells to chemotherapy and thus inhibit cancer progression and recurrence. For example, Kim et al. [38] constructed a cationic liposome DOTAP/DOPE complexed with MALAT1 siRNA to promote the apoptosis of glioblastoma cells during temozolomide treatment. With the aim to cross the blood-brain barrier, the liposome was conjunct with a fragment form, i.e., the transferrin receptor monoclonal antibody. As expected, this nanocomplex reached the cancer lesion and exhibited synergistic effects with temozolomide in decreasing tumor weight. Wu and his colleagues [39] prepared another kind of liposome using DSPE-PEG2000, DSPG, and DSPC and then conjugated it with siRNA targeting the LINC00511. They also transfected the siRNA into breast cancer cells with the help of low-frequency ultrasound (LFUS). They found that cells treated with the nanobubbles in the presence of LFUS exhibited increased chemosensitivity to cisplatin. These innovative approaches may give some inspiration for future nanotargeted therapy. And, with further improvements in the modification of nanostructured carriers, the precise delivery of siRNA to targeted cancer cells may be achieved one day.

Bioactive compounds and Chinese herbal compounds

Today, increasing evidence suggests that TCM not only affects miRNA but also influences lncRNA to unleash anticancer effects. The reversal of chemoresistance effects of compounds isolated from turmeric were mentioned in a previous section. A study conducted by Yoshida's group [40] implied that curcumin could elevate the chemosensitivity of pancreatic cancer to gemcitabine by downregulating the lncRNA PVT1. Mechanically, the anticancer effects were associated with promoting cell apoptosis and cell cycle arrest. In the doxorubicin-resistant osteosarcoma, Polydatin could inactivate the AKT signaling pathway via targeting the lncRNA TUG1, which then facilitated cell apoptosis and inhibited cell proliferation during doxorubicin treatment [41].

Apart from bioactive compounds, Chinese herbal compounds also have been confirmed to regulate lncRNA. Taking the Qingyihuaji formula as an example, it could enhance the anticancer effects of gemcitabine in pancreatic cancer by upregulating the lncRNA AB209630 [42]. During this process, lncRNA AB209630 acted as a ceRNA of miR-373 to active EPHB2/Nanong axis and then affected the functions including cell apoptosis, cell migration, and cell self-renewal. Shenqi Fuzheng injection, a widely used clinical Chinese herbal compound, has been certified to sensitize gastric cells to 5-fluoroura and mitomycin by inhibiting the lncRNA HOTAIR, which targets the miR-17/PTEN signaling pathway to regulate the EMT transition, cell apoptosis, and proliferation [43].

These observations added to the foundation of overcoming chemoresistance by TCM. However, given the multiple target property of TCM, other potential targets of the previously mentioned compounds and compound preparations need to be further illustrated.

The predictive biomarkers

With in-depth understanding and great advancement of cancer management, people today have moved on from the "one approach fits all" strategy to a precise treatment path. Targeted therapy is a representative and well-worked example. Erlotinib, the epidermal growth factor receptor-tyrosine kinase inhibitors (EGFR-TKIs), its indications is NSCLC patients with EGFR mutation instead of all NSCLC patients [44]. From this case, the importance and conspicuous value of predictive biomarkers for cancer treatment are shown.

The dawn and shortage of predictive biomarkers

Some works have been done in identifying and implementing predictive biomarkers for epigenetic therapy. For example, Ivosidenib, the small molecule compound targeting mutant IDH, has shown potent anticancer effects on relapsed or refractory Acute Myeloid Leukemia (AML) patients with IDH mutant [11]. Data of the clinical trial demonstrate that the rate of complete response was up to 20%. Similar phenomenon was observed in a study conducted by Ribrag and his colleagues [45]; malignancy patients accompanied with the EZH2 mutations or SMARCA4 aberrations exhibited potent therapy response to EZH2 inhibitor tazemetostat. In line with these, a clinical study carried out by Stathis' group [46] confirmed the anticancer activity of a bromodomain and extraterminal (BET) inhibitor on NUT midline carcinoma patients, whose pathogenesis was the occurrence of BRD4-NUT fusions. In addition to the previously mentioned studies, there were also some combination therapy clinical trials making efforts to discover potential predictive biomarkers, albeit in a small proportion. A typical case is that data of one phase 1b-2a study indicated that, for ovarian cancer patients suffering platinum resistance, the ones with DR4 hypomethylation might more likely acquire benefits from the combination therapy including platinum and DNMTi decitabine [47].

Albeit with these mentioned findings, it has to be mentioned that there is still a current shortage of predictive biomarkers for epigenetic therapy. This phenomenon may partly contribute to the occurrence of some disappointing outcomes. For example, DOT1L inhibitor exerted potent anticancer activity for treating AML with mixed lineage leukemia rearrangement in preclinical studies, whereas complete remission was only shown in 2 of the total 34 patients in a clinical trial, which was below expected [48].

The factors leading to this shortage of biomarkers are complex and multiple. The following are some potential reasons. First of all, as known, epigenetic regulation is a reversible and dynamic process that could be largely affected by cellular context. Thus, the tumor microenvironment and cell type may be determinant factors for the predictive biomarkers, including genomic changes and epigenetic markers. This may partially explain the discrepancy of the anticancer efficacy of tazemetostat in lymphoma and solid tumors [45]. During an ongoing trial of tazemetostat (NCT02601950), the cohorts of patients were well-designed. And trials of this type may provide some new insights and inspiration in discovering predictive biomarkers.

Moreover, the lack of clinical samples, especially the serial ones, is another obstacle that could not be ignored in identifying the predictive biomarkers. On one side, to ascertain the methylation or acetylation level of target genes, fresh or frozen samples were required. But in practice, there are limited numbers of clinical samples that can be used, and a considerable part of them are formalin-fixed paraffin-embedded tissues [20]. One the other side, cancer is a disease characterized by progression, so measurements of clinical samples from different stages such as pretherapy and posttreatment are required, which is difficult to achieve.

Furthermore, it has to be acknowledged that laboratory models could not completely reflect the clinical scene as cancer is a complex and systemic disease. In this context, it is understandable why no predictive biomarkers of epigenetic therapy have been prospectively verified. Besides the differences of trial design, evaluation standard and detection approaches among studies also make it difficult to identify predictive biomarkers.

The discovering of epigenetic biomarkers

Currently, the advancement of techniques provides more possibilities for discovering and clinically utilizing predictive biomarkers for epigenetic therapy. The main hotspot of this field is the application of liquid biopsies that include cell-free circulating DNA, exosomes, and so on. The exosomes refer to secreted microvesicles that could transport ncRNA to regulate tumor metastasis, resistance, and recurrence [49]. One of the merits of liquid biopsies is that, given the tumor heterogeneity, single-site clinical samples may not reveal the whole molecular character of the lesions, and the genetic and epigenetic alterations within the primary and metastatic pathological tissues could be captured comprehensively via employing liquid biopsies [2]. Other merits of liquid biopsies is that the procedure hardly does harm to patients and the monitoring process can be serial. Practically, the promise of liquid biopsies has already been shown in several reports. For example, using a QIAamp DNA Micro Kit could effectively assess the DNA methylation signature through whole-genome bisulfite sequencing [50]. And next-generation sequencing can be conducted on the miRNA obtained by the commercial miRNA purification kit [51]. With improvements on assessment techniques in liquid biopsy, especially in exosomal, more useful information will soon be available.

Apart from these, there are some other excellent techniques for the discovery of predictive biomarkers. Immunohistochemical staining could identify the methylation, acetylation, and other specific characteristics of clinical tissues such as H3K9me2. RNA sequencing could depict the transcriptional signatures of genomes, which could deliver a message about the expression level of potential biomarkers. Using DNA sequencing to obtain genomic alterations might be an advisable and useful approach, as DNA is considerably stable under most conditions [20]. Besides, the ATAC sequencing has become a preferred approach for studying chromatin accessibility today [52]. Furthermore, having assess at the level of single cells would deepen the knowledge of chromatin heterogeneity and provide opportunities for discovering epigenetic biomarkers [53]. These advances of techniques, along with the in-depth knowledge of epigenetic regulation, may offer more possibilities for treating cancer by an epigenetic approach.

Combination with targeted therapy

The advent and advance of epigenetic agents provides us with a promising weapon for cancer management. Given the essential role of epigenetic regulation in cancer, epigenetic therapy has changed the strategy from monotherapy to combination therapy with hope to improve clinical outcomes. In previous chapters, the combined therapy involving epigenetic agents and cytotoxic drugs to overcome chemoresistance was comprehensively discussed. In this section, the combination of epigenetic agents with targeted therapy will be briefly outlined.

EGFR-TKIs

Today, the emergence of EGFR-TKIs such as gefitinib and erlotinib has enriched cancer management options, especially for NSCLC [44]. Unfortunately, the resistance is severely limited in its clinical utility. Owing to in-depth investigations, it has been confirmed that epigenetic dysregulation was an indispensable machinery accounting for this disappointing phenomenon [54].

It has been indicated that EMT, a process characterized by upregulation of vimentin and downregulation of E-cadherin, was a typical histological transformation leading to EGFR-TKIs resistance [55]. This phenotypic alteration could render cancer cells more invasive and malignant, thereby less sensitive to EGFR-TKIs. Notably, the HDACi vorinostat could reverse the EMT to increase the migration and invasion of gefitinib-resistant cancer cells [56]. The changes then enhanced the anticancer efficiency of gefitinib. Furthermore, another study demonstrated that erlotinib could further decrease cell proliferation when HDAC activity was inhibited, which implied the synergistic effect of HDACi and EGFR-TKIs [57]. The relevant clinical trial was also conducted employing erlotinib and Panobinostat, the HDAC inhibitor [58]. During the study, the disease control rate was 54% and 43% for patients suffering from NSCLC or head-and-neck cancer, respectively. The data also indicated that patients with low level of CHK1 were more likely to obtain benefits from the combined therapy, and the potential of using CHK1 as a predictive biomarker would be further verified in following trials.

What's more, other epigenetic modulators such as HMT and ncRNA were also confirmed to be involved in the process of EGFR-TKIs resistance. Previous work indicated that EHMT2 could be transcriptionally inactivate the PTEN to active AKT, which was a well-known bypass pathway to promote cell survival during EGFR-TKIs treatment [59]. As expected, EHMT2 inhibitor restored the sensitivity of NSCLC to erlotinib and hampered the growth of erlotinib-resistant NSCLC xenograft. Additionally, inhibition of LSD1, the histone demethylase, effectively reversed the resistance of NSCLC to gefitinib under hypoxia circumstance under which condition could promote resistance to EGFR-TKIs [60]. It also has been indicated that miR-483-3p could reverse EMT-mediated gefitinib resistance [61]. Mechanistically, miR-483-3p targeting inhibited integrin β3 to downregulate the FAK/ERK axis.

These observations indeed provided rationale for the combination of EGFR-TKIs and epigenetic agents for cancer treatment. Also, further clinical trials based on these findings may provide more meaningful information like dosage selection, administration schedule, and so on.

Antiangiogenic agents

Tumor angiogenesis as an indispensable part during cancer pathogenesis is a process mainly regulated by hypoxia-inducible factor 1α (HIF1α) and VEGF [62]. It has been demonstrated that abnormal tumor vessels would intensively affect tumor microenvironment, which then influenced cell proliferation, tumor metastasis, and drug delivery [62]. Given the aforementioned, several antiangiogenic agents have been exploited and approved, for example, pazopanib and sorafenib [63].

It noteworthy that several studies have already implied the possibilities of combined antiangiogenic drugs with epigenetic agents. A study executed by Zeng's group [64] indicated that inhibition of EZH2 by GSK126 could attenuate tumor angiogenesis through negatively regulating VEGF-A. Furthermore, EHMT2 inhibitor BIX01294 could significantly attenuate tumor angiogenesis of hepatocellular carcinoma cells under hypoxic condition [65].

The HDACi may be another ideal option. It has been indicated that HDAC2 could maintain HIF1α stability via blocking the interaction between HIF1α and von Hippel–Lindau protein, while HDAC4 may decrease the stability of HIF1α by facilitating its acetylation [66, 67]. Based on these findings, several clinical studies of an early stage were carried out. In a phase I/II study, hepatocellular carcinoma patients obtaining combined therapy including HDACi resminostat and sorafenib have higher PFS-rate and longer OS than patients administrated with resminostat only [68]. Data of the study also indicated that Zinc finger protein 64's level in peripheral blood may be positively correlated with the prognosis. Another study demonstrated that about 70% of renal cell carcinoma patients who received combination therapy involving pazopanib and HDACi abexinostat displayed durable tumor regressions [69]. Besides, the HDAC2 level in peripheral blood was found to be positively associated with the therapy response. These clinical trials confirmed the benefits and tolerance of the combination therapy including HDACi and antiangiogenic agents. The importance and potential of predictive biomarkers were also shown in these studies.

PI3K/AKT/mTOR pathway inhibitors

The PI3K/AKT/mTOR axis is one of the main fundamental signaling pathways for cell survival, which is involved in various cellular functions including cell proliferation, motility, and so on [70]. It also one of the most common activated axis in cancer cells [70]. Given its versatile properties and essential role in cancer progression, drugs targeting PI3K/AKT/mTOR axis were exploited to enrich clinical options.

Furthermore, multiple findings have implied the possibilities of combined epigenetic agents with inhibitors of PI3K/AKT/mTOR axis to improve anticancer efficacy. For example, it has been indicated that inhibition of HDAC1 and HDAC2 could reduce malignance and stemness of cancer cells partially through deactivating the PI3K/AKT/mTOR axis [71]. Besides, HDACi has been reported to cooperate with PI3K inhibitor in suppressing the growth of medulloblastoma xenograft [72]. Mechanically, they were synergetic to facilitate the activation of FOXO1. Inspired by these, CUDC-907, the HDAC and PI3K dual inhibitor, was designed and synthesized [72]. As expected, it exhibited effective anticancer activity in the PDX model of prostate cancer and AML [73, 74], and its anticancer effects were correlated with facilitating DNA damage by downregulating CHK1, RRM1, PRM2, and Wee1.

Additionally, inhibitors targeting BRD4 also displayed synergistic effects with PI3K/AKT/mTOR axis inhibitors in related studies. It has been reported that targeting at BRD4 could restore the sensitivity of colorectal cancer cells to PI3K/mTOR dual inhibitors and then induce cell death [74]. Furthermore, similar to CUDC-907, SF2523 is a dual inhibitor targeting at PI3K and BRD4 that displayed intense anticancer effects. The work conducted by Kutateladze et al. [75] suggested that SF2523 could effectively suppress growth and metastasis of cancer cells in vivo through downregulating the MYC. Another study demonstrated that SF2523 could also delay tumor growth by preventing tumor immunosuppression induced by macrophages [76].

These mentioned studies offered the rationale for combined epigenetic agents with PI3K/AKT/mTOR axis inhibitors. They also shed insights on the discovery and design of dual targets inhibitors, which affecting both epigenetic regulator and PI3K/AKT/mTOR pathway.

Conclusion

The advancement of epigenetic agents provides us a promising pharmacological weapon for cancer management. Furthermore, elaboration of the role of epigenetic regulation in chemoresistance offers opportunities to sensitize cancer cells to chemotherapy by targeting this machinery. As expected, some encouraging outcomes were achieved using agents acting at various epigenetic regulators involving DNMT, HDAC, miRNA, and so on. However, there were still some disappointing results, which may be improved by a more rational administration schedule, lower dosage, longer treatment duration, and in-depth knowledge of the role of epigenetic machinery in chemoresistance. The disappointing results also remind us that epigenetic therapy is not panacea that can solve all issues once and for all. Just as targeted therapy is a precision-medicine treatment that selects target patients by predictive biomarkers, epigenetic therapy might be specifically effective for some subclasses of patients.

Following this concept, increasing studies recruited patients based on certain predictive biomarkers and acquired positive results. This innovative approach may provide more possibilities for epigenetic therapy. What's more, in addition to the involvement in chemoresistance, epigenetic regulation also has been confirmed to synergy with targeted therapy such as EGFR-TKIs and antiangiogenic agents. This may extend the clinical applications of epigenetic therapy in the near future. In brief, we are optimistic that the complete potential of epigenetic therapy will be achieved one day by thoroughly illustrating the properties of epigenetic regulation and its role in cancer progression. Epigenetic therapy may ultimately provide more powerful clinical options for cancer management.

References

[1] Lu Y, Chan YT, Tan HY, Li S, Wang N, Feng Y. Epigenetic regulation in human cancer: the potential role of epi-drug in cancer therapy. Mol Cancer 2020;19(1):79.
[2] Chatterjee N, Bivona TG. Polytherapy and targeted cancer drug resistance. Trends Cancer 2019;5(3):170–82.
[3] Wang L, Liu X, Ren Y, Zhang J, Chen J, Zhou W, et al. Cisplatin-enriching cancer stem cells confer multidrug resistance in non-small cell lung cancer via enhancing TRIB1/HDAC activity. Cell Death Dis 2017;8(4), e2746.
[4] Dawson MA. The cancer epigenome: concepts, challenges, and therapeutic opportunities. Science 2017;355 (6330):1147–52.
[5] Kelly AD, Issa JJ. The promise of epigenetic therapy: reprogramming the cancer epigenome. Curr Opin Genet Dev 2017;42:68–77.
[6] Miranda Furtado CL, Dos Santos Luciano MC, Silva Santos RD, Furtado GP, Moraes MO, Pessoa C. Epidrugs: targeting epigenetic marks in cancer treatment. Epigenetics 2019;14(12):1164–76.
[7] Jones PA, Issa JP, Baylin S. Targeting the cancer epigenome for therapy. Nat Rev Genet 2016;17(10):630–41.
[8] Hoy SM. Tazemetostat: first approval. Drugs 2020;80(5):513–21.
[9] Ramalingam SS, Maitland ML, Frankel P, Argiris AE, Koczywas M, Gitlitz B, et al. Carboplatin and paclitaxel in combination with either vorinostat or placebo for first-line therapy of advanced non-small-cell lung cancer. J Clin Oncol 2010;28(1):56–62.
[10] Appleton K, Mackay HJ, Judson I, Plumb JA, McCormick C, Strathdee G, et al. Phase I and pharmacodynamic trial of the DNA methyltransferase inhibitor decitabine and carboplatin in solid tumors. J Clin Oncol 2007; 25(29):4603–9.
[11] DiNardo CD, Stein EM, de Botton S, Roboz GJ, Altman JK, Mims AS, et al. Durable remissions with Ivosidenib in IDH1-mutated relapsed or refractory AML. N Engl J Med 2018;378(25):2386–98.
[12] Azad N, Zahnow CA, Rudin CM, Baylin SB. The future of epigenetic therapy in solid tumours- -lessons from the past. Nat Rev Clin Oncol 2013;10(5):256–66.
[13] Fakih MG, Fetterly G, Egorin MJ, Muindi JR, Espinoza-Delgado I, Zwiebel JA, et al. A phase I, pharmacokinetic, and pharmacodynamic study of two schedules of vorinostat in combination with 5-fluorouracil and leucovorin in patients with refractory solid tumors. Clin Cancer Res 2010;16(14):3786–94.
[14] Jiao X, Qian X, Wu L, Li B, Wang Y, Kong X, et al. microRNA: the impact on cancer stemness and therapeutic resistance. Cells 2019;9(1).
[15] He B, Zhao Z, Cai Q, Zhang Y, Zhang P, Shi S, et al. miRNA-based biomarkers, therapies, and resistance in Cancer. Int J Biol Sci 2020;16(14):2628–47.
[16] Huang BS, Luo QZ, Han Y, Huang D, Tang QP, Wu LX. MiR-223/PAX6 axis regulates glioblastoma stem cell proliferation and the chemo resistance to TMZ via regulating PI3K/Akt pathway. J Cell Biochem 2017; 118(10):3452–61.
[17] Zheng Y, Lv X, Wang X, Wang B, Shao X, Huang Y, et al. MiR-181b promotes chemoresistance in breast cancer by regulating Bim expression. Oncol Rep 2016;35(2):683–90.
[18] Sun H, Zhou X, Bao Y, Xiong G, Cui Y, Zhou H. Involvement of miR-4262 in paclitaxel resistance through the regulation of PTEN in non-small cell lung cancer. Open Biol 2019;9(7):180227.

[19] Zhou Y, Chen E, Tang Y, Mao J, Shen J, Zheng X, et al. miR-223 overexpression inhibits doxorubicin-induced autophagy by targeting FOXO3a and reverses chemoresistance in hepatocellular carcinoma cells. Cell Death Dis 2019;10(11):843.

[20] Morel D, Jeffery D, Aspeslagh S, Almouzni G, Postel-Vinay S. Combining epigenetic drugs with other therapies for solid tumours - past lessons and future promise. Nat Rev Clin Oncol 2020;17(2):91–107.

[21] Wang H, Liu S, Jia L, Chu F, Zhou Y, He Z, et al. Nanostructured lipid carriers for MicroRNA delivery in tumor gene therapy. Cancer Cell Int 2018;18:101.

[22] Desantis V, Saltarella I, Lamanuzzi A, Melaccio A, Solimando AG, Mariggio MA, et al. MicroRNAs-based nano-strategies as new therapeutic approach in multiple myeloma to overcome disease progression and drug resistance. Int J Mol Sci 2020;21(9).

[23] Liu J, Meng T, Yuan M, Wen L, Cheng B, Liu N, et al. MicroRNA-200c delivered by solid lipid nanoparticles enhances the effect of paclitaxel on breast cancer stem cell. Int J Nanomedicine 2016;11:6713–25.

[24] Xie Y, Wang Y, Li J, Hang Y, Jaramillo L, Wehrkamp CJ, et al. Cholangiocarcinoma therapy with nanoparticles that combine downregulation of MicroRNA-210 with inhibition of cancer cell invasiveness. Theranostics 2018; 8(16):4305–20.

[25] Liang G, Zhu Y, Ali DJ, Tian T, Xu H, Si K, et al. Engineered exosomes for targeted co-delivery of miR-21 inhibitor and chemotherapeutics to reverse drug resistance in colon cancer. J Nanobiotechnol 2020;18(1):10.

[26] Xiang Y, Guo Z, Zhu P, Chen J, Huang Y. Traditional Chinese medicine as a cancer treatment: modern perspectives of ancient but advanced science. Cancer Med 2019;8(5):1958–75.

[27] Zeng C, Fan D, Xu Y, Li X, Yuan J, Yang Q, et al. Curcumol enhances the sensitivity of doxorubicin in triple-negative breast cancer via regulating the miR-181b-2-3p-ABCC3 axis. Biochem Pharmacol 2020;174:113795.

[28] Yu C, Chen DQ, Liu HX, Li WB, Lu JW, Feng JF. Rosmarinic acid reduces the resistance of gastric carcinoma cells to 5-fluorouracil by downregulating FOXO4-targeting miR-6785-5p. Biomed Pharmacother 2019;109:2327–34.

[29] Fu W, Hong Z, You X, Din J, Chen B, Zhao B, et al. Enhancement of anticancer activity of docetaxel by combination with Fuzheng Yiliu decoction in a mouse model of castration-resistant prostate cancer. Biomed Pharmacother 2019;118:109374.

[30] Bach DH, Lee SK. Long noncoding RNAs in cancer cells. Cancer Lett 2018;419:152–66.

[31] Wang Y, Fang Z, Hong M, Yang D, Xie W. Long-noncoding RNAs (lncRNAs) in drug metabolism and disposition, implications in cancer chemo-resistance. Acta Pharm Sin B 2020;10(1):105–12.

[32] Li Z, Niu H, Qin Q, Yang S, Wang Q, Yu C, et al. lncRNA UCA1 mediates resistance to cisplatin by regulating the miR-143/FOSL2-signaling pathway in ovarian cancer. Mol Ther Nucleic Acids 2019;17:92–101.

[33] Liu B, Wu S, Ma J, Yan S, Xiao Z, Wan L, et al. lncRNA GAS5 reverses EMT and tumor stem cell-mediated gemcitabine resistance and metastasis by targeting miR-221/SOCS3 in pancreatic cancer. Mol Ther Nucleic Acids 2018;13:472–82.

[34] Ye Y, Yang S, Han Y, Sun J, Xv L, Wu L, et al. HOXD-AS1 confers cisplatin resistance in gastric cancer through epigenetically silencing PDCD4 via recruiting EZH2. Open Biol 2019;9(9):190068.

[35] Wu H, Gu J, Zhou D, Cheng W, Wang Y, Wang Q, et al. LINC00160 mediated paclitaxel-and doxorubicin-resistance in breast cancer cells by regulating TFF3 via transcription factor C/EBPbeta. J Cell Mol Med 2020;24(15):8589–602.

[36] Xu X, Zhou X, Chen Z, Gao C, Zhao L, Cui Y. Silencing of lncRNA XIST inhibits non-small cell lung cancer growth and promotes chemosensitivity to cisplatin. Aging 2020;12(6):4711–26.

[37] Cai Q, Wang S, Jin L, Weng M, Zhou D, Wang J, et al. Long non-coding RNA GBCDRlnc1 induces chemoresistance of gallbladder cancer cells by activating autophagy. Mol Cancer 2019;18(1):82.

[38] Kim SS, Harford JB, Moghe M, Rait A, Pirollo KF, Chang EH. Targeted nanocomplex carrying siRNA against MALAT1 sensitizes glioblastoma to temozolomide. Nucleic Acids Res 2018;46(3):1424–40.

[39] Wu B, Yuan Y, Han X, Wang Q, Shang H, Liang X, et al. Structure of LINC00511-siRNA-conjugated nanobubbles and improvement of cisplatin sensitivity on triple negative breast cancer. FASEB J 2020;34(7):9713–26.

[40] Yoshida K, Toden S, Ravindranathan P, Han H, Goel A. Curcumin sensitizes pancreatic cancer cells to gemcitabine by attenuating PRC2 subunit EZH2, and the lncRNA PVT1 expression. Carcinogenesis 2017; 38(10):1036–46.

[41] Hu T, Fei Z, Su H, Xie R, Chen L. Polydatin inhibits proliferation and promotes apoptosis of doxorubicin-resistant osteosarcoma through LncRNA TUG1 mediated suppression of Akt signaling. Toxicol Appl Pharmacol 2019;371:55–62.

[42] Chen P, Wang M, Wang C. Qingyihuaji formula reverses gemcitabine resistant human pancreatic cancer through regulate lncRNA AB209630/miR-373/EphB2-NANOG signals. Biosci Rep 2019;39(6).
[43] Jia J, Zhan D, Li J, Li Z, Li H, Qian J. The contrary functions of lncRNA HOTAIR/miR-17-5p/PTEN axis and Shenqifuzheng injection on chemosensitivity of gastric cancer cells. J Cell Mol Med 2019;23(1):656–69.
[44] Rotow J, Bivona TG. Understanding and targeting resistance mechanisms in NSCLC. Nat Rev Cancer 2017; 17(11):637–58.
[45] Italiano A, Soria JC, Toulmonde M, Michot JM, Lucchesi C, Varga A, et al. Tazemetostat, an EZH2 inhibitor, in relapsed or refractory B-cell non-Hodgkin lymphoma and advanced solid tumours: a first-in-human, open-label, phase 1 study. Lancet Oncol 2018;19(5):649–59.
[46] Stathis A, Zucca E, Bekradda M, Gomez-Roca C, Delord JP, de La Motte RT, et al. Clinical response of carcinomas harboring the BRD4-NUT oncoprotein to the targeted bromodomain inhibitor OTX015/MK-8628. Cancer Discov 2016;6(5):492–500.
[47] Fu S, Hu W, Iyer R, Kavanagh JJ, Coleman RL, Levenback CF, et al. Phase 1b-2a study to reverse platinum resistance through use of a hypomethylating agent, azacitidine, in patients with platinum-resistant or platinum-refractory epithelial ovarian cancer. Cancer 2011;117(8):1661–9.
[48] Stein EM, Tallman MS. Emerging therapeutic drugs for AML. Blood 2016;127(1):71–8.
[49] Xie Y, Dang W, Zhang S, Yue W, Yang L, Zhai X, et al. The role of exosomal noncoding RNAs in cancer. Mol Cancer 2019;18(1):37.
[50] Legendre C, Gooden GC, Johnson K, Martinez RA, Liang WS, Salhia B. Whole-genome bisulfite sequencing of cell-free DNA identifies signature associated with metastatic breast cancer. Clin Epigenetics 2015;7:100.
[51] Seco-Cervera M, Gonzalez-Rodriguez D, Ibanez-Cabellos JS, Peiro-Chova L, Gonzalez-Cabo P, Garcia-Lopez E, et al. Circulating miR-323-3p is a biomarker for cardiomyopathy and an indicator of phenotypic variability in Friedreich's ataxia patients. Sci Rep 2017;7(1):5237.
[52] Garcia-Gimenez JL, Seco-Cervera M, Tollefsbol TO, Roma-Mateo C, Peiro-Chova L, Lapunzina P, et al. Epigenetic biomarkers: current strategies and future challenges for their use in the clinical laboratory. Crit Rev Clin Lab Sci 2017;54(7–8):529–50.
[53] Grosselin K, Durand A, Marsolier J, Poitou A, Marangoni E, Nemati F, et al. High-throughput single-cell ChIP-seq identifies heterogeneity of chromatin states in breast cancer. Nat Genet 2019;51(6):1060–6.
[54] Rosenzweig SA. Acquired resistance to drugs targeting tyrosine kinases. Adv Cancer Res 2018;138:71–98.
[55] Weng CH, Chen LY, Lin YC, Shih JY, Lin YC, Tseng RY, et al. Epithelial-mesenchymal transition (EMT) beyond EGFR mutations per se is a common mechanism for acquired resistance to EGFR TKI. Oncogene 2019;38(4):455–68.
[56] Bruzzese F, Leone A, Rocco M, Carbone C, Piro G, Caraglia M, et al. HDAC inhibitor vorinostat enhances the antitumor effect of gefitinib in squamous cell carcinoma of head and neck by modulating ErbB receptor expression and reverting EMT. J Cell Physiol 2011;226(9):2378–90.
[57] Liffers K, Kolbe K, Westphal M, Lamszus K, Schulte A. Histone deacetylase inhibitors resensitize EGFR/EGFRvIII-overexpressing, Erlotinib-resistant glioblastoma cells to tyrosine kinase inhibition. Target Oncol 2016;11(1):29–40.
[58] Gray JE, Haura E, Chiappori A, Tanvetyanon T, Williams CC, Pinder-Schenck M, et al. A phase I, pharmacokinetic, and pharmacodynamic study of panobinostat, an HDAC inhibitor, combined with erlotinib in patients with advanced aerodigestive tract tumors. Clin Cancer Res 2014;20(6):1644–55.
[59] Wang L, Dong X, Ren Y, Luo J, Liu P, Su D, et al. Targeting EHMT2 reverses EGFR-TKI resistance in NSCLC by epigenetically regulating the PTEN/AKT signaling pathway. Cell Death Dis 2018;9(2):129.
[60] Lu Y, Liu Y, Oeck S, Glazer PM. Hypoxia promotes resistance to EGFR inhibition in NSCLC cells via the histone demethylases, LSD1 and PLU-1. Mol Cancer Res 2018;16(10):1458–69.
[61] Yue J, Lv D, Wang C, Li L, Zhao Q, Chen H, et al. Epigenetic silencing of miR-483-3p promotes acquired gefitinib resistance and EMT in EGFR-mutant NSCLC by targeting integrin beta3. Oncogene 2018;37(31):4300–12.
[62] Abdalla AME, Xiao L, Ullah MW, Yu M, Ouyang C, Yang G. Current challenges of cancer anti-angiogenic therapy and the promise of nanotherapeutics. Theranostics 2018;8(2):533–48.
[63] Potente M, Gerhardt H, Carmeliet P. Basic and therapeutic aspects of angiogenesis. Cell 2011;146(6):873–87.
[64] Chen YT, Zhu F, Lin WR, Ying RB, Yang YP, Zeng LH. The novel EZH2 inhibitor, GSK126, suppresses cell migration and angiogenesis via down-regulating VEGF-A. Cancer Chemother Pharmacol 2016;77(4):757–65.
[65] Oh SY, Seok JY, Choi YS, Lee SH, Bae JS, Lee YM. The histone methyltransferase inhibitor BIX01294 inhibits HIF-1alpha stability and angiogenesis. Mol Cells 2015;38(6):528–34.

[66] Chang CC, Lin BR, Chen ST, Hsieh TH, Li YJ, Kuo MY. HDAC2 promotes cell migration/invasion abilities through HIF-1alpha stabilization in human oral squamous cell carcinoma. J Oral Pathol Med 2011;40(7):567–75.
[67] Geng H, Harvey CT, Pittsenbarger J, Liu Q, Beer TM, Xue C, et al. HDAC4 protein regulates HIF1alpha protein lysine acetylation and cancer cell response to hypoxia. J Biol Chem 2011;286(44):38095–102.
[68] Bitzer M, Horger M, Giannini EG, Ganten TM, Worns MA, Siveke JT, et al. Resminostat plus sorafenib as second-line therapy of advanced hepatocellular carcinoma - the SHELTER study. J Hepatol 2016;65(2):280–8.
[69] Aggarwal R, Thomas S, Pawlowska N, Bartelink I, Grabowsky J, Jahan T, et al. Inhibiting histone deacetylase as a means to reverse resistance to angiogenesis inhibitors: phase I study of Abexinostat plus Pazopanib in advanced solid tumor malignancies. J Clin Oncol 2017;35(11):1231–9.
[70] Alzahrani AS. PI3K/Akt/mTOR inhibitors in cancer: at the bench and bedside. Semin Cancer Biol 2019;59:125–32 [Epub 2019/07/20].
[71] Ma S, Liu T, Xu L, Wang Y, Zhou J, Huang T, et al. Histone deacetylases inhibitor MS-275 suppresses human esophageal squamous cell carcinoma cell growth and progression via the PI3K/Akt/mTOR pathway. J Cell Physiol 2019;234(12):22400–10 [Epub 2019/05/24].
[72] Pei Y, Liu KW, Wang J, Garancher A, Tao R, Esparza LA, et al. HDAC and PI3K antagonists cooperate to inhibit growth of MYC-driven medulloblastoma. Cancer Cell 2016;29(3):311–23 [Epub 2016/03/16].
[73] Hu C, Xia H, Bai S, Zhao J, Edwards H, Li X, et al. CUDC-907, a novel dual PI3K and HDAC inhibitor, in prostate cancer: antitumour activity and molecular mechanism of action. J Cell Mol Med 2020; [Epub 2020/05/28].
[74] Li X, Su Y, Madlambayan G, Edwards H, Polin L, Kushner J, et al. Antileukemic activity and mechanism of action of the novel PI3K and histone deacetylase dual inhibitor CUDC-907 in acute myeloid leukemia. Haematologica 2019;104(11):2225–40 [Epub 2019/03/02].
[75] Andrews FH, Singh AR, Joshi S, Smith CA, Morales GA, Garlich JR, et al. Dual-activity PI3K-BRD4 inhibitor for the orthogonal inhibition of MYC to block tumor growth and metastasis. Proc Natl Acad Sci U S A 2017;114(7): E1072–80 [Epub 2017/02/01].
[76] Joshi S, Singh AR, Liu KX, Pham TV, Zulcic M, Skola D, et al. SF2523: dual PI3K/BRD4 inhibitor blocks tumor immunosuppression and promotes adaptive immune responses in Cancer. Mol Cancer Ther 2019;18(6):1036–44 [Epub 2019/04/26].

Index

Note: Page numbers followed by *f* indicate figures and *t* indicate tables.

FSC
www.fsc.org
MIX
Papier | Fördert gute Waldnutzung
FSC® C083411

Kolibri 360 GmbH
Ferdinand-Jühlke-Straße 7
99095 Erfurt, Deutschland
produktsicherheit@kolibri360.de